I0053525

Frontiers in Occupational Health and Safety

Changes in the World of Work and Impacts on Occupational Health and Safety

Volume 1

Editor

Marcela G. Ribeiro

Fundação Jorge Duprat Figueiredo de Segurança e Medicina do Trabalho - FUNDACENTRO
São Paulo
Brazil

CONTENTS

FOREWORD

The quick global changes that we witnessed in the last two decades caused a profound impact in the world of work, both in highly industrialized societies and in countries in different phases of development. *Frontiers in Occupational Health and Safety* starts its series with an in depth discussion of work organization facing the present complexities, new technologies exemplified by the rapid surge of nanotechnology, and an interesting analysis of the competencies and required profiles of Occupational Safety and Health professionals, necessary to tackle these new challenges. This first volume carefully edited by Dr. Marcela Gerardo Ribeiro is an instigating beginning, setting the floor for the upcoming volumes of this series.

Eduardo Algranti, MD, PhD
Division of Medicine
Fundacentro
Brazil

PREFACE

This is the first issue of the upcoming series entitled **Frontiers of Occupational Health and Safety**. The first volume, *Changes in the World of Work and Impacts on Occupational Health and Safety*, will focus on certain aspects of how the global economic, political, and cultural changes have reshaped the world of work.

The book begins with a discussion on the changes in work organization and their impacts on occupational safety and health. Some of the main aspects of new forms of work organization (NFWO) are presented. Dias and Lima propose that within NFWO there may be some paradoxes that workers are subjected to deal with. As a consequence of the performance control present in management systems of global organizations, as a NFWO, new work pathologies emerge, psychological ones, ordinarily named *stress*.

As a consequence of the fast technological development within processes and products, new risks emerge in the workplace. In such context, Martins and Dulley present a broad reflection of the impacts of new risks into workers' universe, using nanotechnology as a backdrop. Nanotechnology encompasses an extraordinary diversity of technological approaches currently under development. This has created a relatively new industry. The goods being produced and introduced into the market are usually innovative and unfamiliar to most consumers. It was identified as an important source of known and unknown risks, and due to the increasing uncertainties posed by these technologies, regarding unique challenges in occupational health and safety, as well as environmental, legal, societal and ethical impacts.

At last, the changes in employment relationship; workforce demographics; workplace environment; and work organization, among others, have profoundly affected not only the working life, but also the structure of Occupational Health & Safety (OHS) professions, and the role of OHS professionals within organizations, workplace, workforce, and society. OHS professionals are and will be a necessary part of doing business worldwide, as researchers, educators, or practitioners. Ribeiro and Ventura seek to point out the concerning needs on OHS professional competencies demanded by such changes. Rather than giving clear and finite answers to such challenging questions, the authors invite reflections on what are at stake regarding OHS professionals training and education, skills and research.

There is much yet to be discussed regarding the frontiers of occupational health and safety. This book gives the reader hints of the complex and broad world of work universe.

Marcela G. Ribeiro, PhD
Fundacentro
São Paulo
Brazil

LIST OF CONTRIBUTORS

Dias, A. V. C.

Department of Production Engineering, Federal University of Minas Gerais, Belo Horizonte, MG, Brazil

Dulley, R. D.

Brazilian Research Network in Nanotechnology, Society and Environment Renanosoma, São Paulo, SP, Brazil

Lima, F. P. A.

Department of Production Engineering, Federal University of Minas Gerais, Belo Horizonte, MG, Brazil

Martins, P. R.

Brazilian Research Network in Nanotechnology, Society and Environment Renanosoma, São Paulo, SP, Brazil

Ribeiro, M. G.

Fundação Jorge Duprat Figueiredo de Segurança e Medicina do Trabalho – Fundacentro, São Paulo, SP, Brazil

Ventura, F. F.

Fundação Jorge Duprat Figueiredo de Segurança e Medicina do Trabalho – Fundacentro, São Paulo, SP, Brazil

CHAPTER 1

Work Organization and Occupational Health in Contemporary Capitalism

Ana Valéria C. Dias[*] and Francisco de P. A. Lima

Department of Production Engineering, Federal University of Minas Gerais, Belo Horizonte, MG, Brazil

Abstract: This chapter discusses recent changes in work organization and their impacts on occupational health and safety. In the last decades of the 20th century, changes in the technological, social and economic context, such as automation of production processes, globalization of markets, financialization and new social demands fostered the emergence of new rationales for production and as a consequence new rationales of work organization. Some of the main aspects of New Forms of Work Organization (NFWO) are presented: flexibility, autonomy, the importance of workers' competence and engagement and management by goals, represented by Key Performance Indicators (KPIs). We then propose that within NFWO there may be some paradoxes that workers are obliged to deal with. Beyond NFWO, some new configurations of firms, notably those based on network features, are also presented, and their relation with NFWO is discussed. In this new context organization, there are new work pathologies, namely the psychological ones, ordinarily named "stress". Final discussion points to the fact that the absence of prescribed tasks created new constraints that are behind the bullying at work. The actual augmentation of psychological suffering and mental diseases is the consequence of the performance control through KPIs that are present in management systems diffused in the corporate governance of global organizations and networks.

Keywords: Automation, bullying, competence, global productive chains, global productive networks, hazards exportation, immaterial work, international division of labor, key performance indicators, occupational health and safety, peripheral countries, stress, worker engagement, work ergonomic analysis, work organization.

INTRODUCTION

The impact of work organization over the health of workers has been recognized for a long time. Indeed, it can be stated that the very nature of so-called French

Corresponding author Ana Valéria C. Dias: Department of Production Engineering, Federal University of Minas Gerais, Belo Horizonte, MG, Brazil; E-mail: anaval@ufmg.br

Marcela G. Ribeiro (Ed)
All rights reserved-© 2014 Bentham Science Publishers

Ergonomics, or Activity Ergonomics, is to discuss and, to some extent, confront some of the basic assumptions of classical work organization forms – namely, Taylorism and Fordism – such as the need for prescriptions and the selection of worker according to the prescribed task. The Work Ergonomic Analysis (WEA) has been able to demonstrate that the real activity performed by workers is different from the prescribed task, and that diseases may appear if this difference is not recognized. In this gap between task and real activity, the workers deployed different regulations strategies, variable according the contingents work situations. Eventually, these regulations overleap the limits of acceptable workload, generating pathogenic effects.

The opposition 'real activity versus prescribed task' has therefore been the center of many ergonomic inquiries during the last decades of the 20th century, corroborated by researches in others fields like sociology and psychology of work, all of them critical about the pathogenic effects of the work rationalization. Nevertheless, technological, social and economic changes that have been in course since at least 1960 provoked changes in work organization. These changes were modest at first, but proved to be dramatic in the beginning of the 2000s. For instance, work time flexibility and the possibility of work at home are wide spread practices in some industries, for instance, the software industry. May that mean that control over workers has decreased? Without rigidly prescribed times and movements, the workers have more autonomy to regulate the workload? Are workers of industries such as the software one less subject to occupational diseases?

Take the work time flexibility as an example. At Google, for instance, work time schedules are flexible and a well-known practice is that 20 percent of the work time may be dedicated to the development of new ideas, or as Google calls it 'pet ideas' i.e. ideas that the software developer may have had by him or herself, and that have not been appreciated by none of his or hers superiors or colleagues (Savoia & Copeland, 2011). In fact, there is no prescription of how to come up with an idea, or how to develop it up to the point of presenting it to someone else. Thus in this case there is no such opposition as prescribed task versus real activity. Does that signify than developers at Google are completely free at work and protected from occupational related health damages? Is the classical work organization framework suitable to this case? How should researchers and practitioners who aim at investigating workers' health approach a work reality such as that at the software industry in general?

The aim of this chapter is to discuss new forms of work organization (NFWO) and how they affect workers. The chapter is organized as follows. After a brief presentation of some main aspects of Taylorism-Fordism, we examine the new rationales of NFWO considering their relationship with technological, social and economic context. The rising importance of immaterial work for contemporary capitalism is also discussed, in particular considering automated production processes. We then propose that within NFWO there may be some paradoxes that workers are obliged to deal with. Beyond NFWO, some new configurations of firms, notably those based on network features, are also presented, and their relation with NFWO is discussed. In this new context organization, there are new work pathologies, namely the psychological ones, ordinarily named '*stress*'. We will see that the absence of prescribed tasks created new constraints that are behind the bulling at work. The actual augmentation of psychological suffering and mental diseases is the consequence of the performance control through key performance indicators (KPI) that are present in management systems diffused in the corporate governance of global organizations and networks.

CHANGING THE CONTEXT: THE LIMITS OF TAYLORISM-FORDISM AND THE EMERGENCE OF NEW FORMS OF WORK ORGANIZATION

The classic forms of work organization, mainly based on the work of Frederick W. Taylor and Henry Ford, have dominated industrial life throughout the 20th century; however, some of their principles have recently been object of interrogation as they would not fit economic, technological and social reality of the late 20th and early 21st centuries. Taylor's Scientific Management theory was built over three principles: the need for a 'scientific analysis of work'; the 'scientific' selection and training of workers according to the task once prescribed; and the daily planning and control of work by management. In brief, Taylor divorced workers from work design, and created (or aimed at creating) an 'abstract' work, or task, which should be prescribed independently of workers. In fact, the development of the task is the first principle, while the selection of workers is the second one; in other words, task precedes worker and therefore its existence is self-contained. According to Taylor, it is always possible to describe, analyze and prescribe work through concrete rules and procedures in order to introduce management control in the shop floor, but also aiming at introducing science in the shop floor, transforming work into a predictable element that could be optimized to increase performance and profits. It is important to note that both first and third principles contributed, in a way, to make labor process more 'transparent' to the management and the investors, as they eliminated the 'opacity' that was present when the worker himself dominated the process technology and knowledge. These 'professional secrets' are, in fact, the problem to be solved by the scientific analysis, a sufficient condition to define objectively a rational workload or 'normal workday'.

By its turn, Ford essentially increased the degree of specialization or horizontal division of work, by means of the moving assembly line that intended to increase the rate of production (quantity per hour). The rhythm of the work is paced by the conveyor. Time is the main criteria for productivity of work, as it was crucial to increase quantity within a given period of time, in a context of economies of scale (Zarifian, 1999; Veltz, 2000).

As social, economic and technological contexts changed, these principles have increasingly been criticized. First of all, work was never completely liable of objective description or concrete prescription, as the French Ergonomics and its twin concepts of 'task' ('prescribed work') and 'activity' (actual work) have largely discussed and shown (Wisner, 1995; Guérin *et al.*, 2006; Keiser, 1991; Daniellou, 2005; Daniellou & Rabardel, 2005). The introduction of automation in production processes has changed the nature of work itself. Work is no longer a – up to a point – visible combination of gestures, but it is essentially to monitor the variables and set points of a process, and to promptly act to hinder or correct deviations when they are about to happen. Some attempts to '*taylorize*' this new work points to the way of indicating to the worker which action should follow each possible deviation. Even so, at least two problems remain to be solved by tayloristic approaches: first, these types of procedures are of course only possible to be prescribed if the deviation has already happen at least once; second, it continues to be impossible to prescribe how worker must monitor the process to anticipate deviations and how cognitive work (signal interpretations, diagnostic and decisions) should happen. When we try to specify how the operator in control room must act in order to regulate the process, we fall in an infinite regress due the impossibility to know all the singular combinations in a complex production system. A similar reasoning may be applied to the introduction of information technology into administrative or service processes; according to Gorz (2005), in this case to work is to manage information flows and social interactions. In order to accomplish this goal, workers must mobilize their knowledge and creative reasoning in non-standards situations.

Finally, changes in the economic context, more specifically the competitive context, raise concern about the emphasis on economies of scale as the main source of productivity of capital. As competition increases, low prices by means of large volumes of undiversified products are insufficient to generate profits. In order to differentiate itself from competitors, issues such as product and process quality became important; increased product or service variety and lower product lifecycles call for flexibility and innovation processes within production systems. The workers adaptability and responsively (and responsibility or commitment) is necessary to attend the demands of a relatively changeable and unforeseeable context. Another way to differentiation is to add new functionalities to the product or service, improving the 'service level' of each product or service, i.e. adding value that is perceived by the consumer (or citizen, in the case of a public service). In this conception, to deliver 'service' means to conceive and make a product or service that will fit the needs of the consumer or citizen (Gadrey & Zarifian, 2002). In order to do this, it is necessary for the worker to understand which those needs are; therefore, to work is also to manage social interactions (Zarifian, 2002). All these possibilities show that, in a way, competitiveness depends less on material work and more on "immaterial work" (Gorz, 2005). Indeed, it is through 'immaterial work' that new service and products are thought up; that workers' engagement to perform a good service relation is produced; that workers' competence to optimize production processes is put in practice. In fact, all work is immaterial: the skilled artisan hand is historical and cultural 'tool', so, a corporal technology informed by social and cultural norms that are essentially immaterial. The skill is not a physiological movement or force, but an educated gesture. The most important difference here, with services, is that language is the immediate tool and the medium of the interactions between workers and users.

This new panorama increasingly demanded new forms of work organization. The quest for product and process quality and product flexibility favored the adoption of some aspects of the so-called Toyota Production System (TPS), or lean production. In addition to changes in the way production is scheduled, reducing inventory through just-in-time and pulled production techniques, work organization also was transformed, as workers became multifunctional, performing different tasks such as production, maintenance, housekeeping and quality control (Coriat, 2000). Workers are also organized in teams, assigned to production cells. Within these cells they rotate productive functions. Workers must also collaborate with quality and cost reduction by suggesting improvements to production process, through kaizen (continuous improvement) programs. These suggestions may be thought up individually or within groups. Notwithstanding the fact that within TPS workers have the right and the duty to intervene whenever product or process quality are in danger (Coriat, 2000), standard operations procedures are still important as a control mechanism.

On the other hand, the discussion on new forms of work organization that would have as main feature the autonomy to direct workers, as opposed to the normative-prescriptive character of Taylorist-Fordist organization goes back to the 1950s/1960s with the work of the Tavistock Institute and the development of an alternative, socio-technical model of work organization. The Socio-Technical Systems (STS) aimed at re-designing production systems through joint optimisation of technical and social subsystems, with a strong emphasis in workers' participation (Cherns, 1987). The result of a STS design process would be a democratic organisation, where technical and economic goals should be achieved together with human goals (Eijnateen, 1993). Perhaps the main practical feature of a STS is the teamwork achieved by the semi-autonomous groups, also known as self-management teams. Within semi-autonomous groups, workers are multifunctional, i.e. must perform different tasks and may have autonomy to decide on issues related to production levels, production procedures, quality, sequence and division of tasks, maintenance, and budget, among others. There are fewer hierarchical levels in a socio-technical organization, and management control over workers from the semi-autonomous groups is mainly accomplished by means of KPI. Table **1** shows some characteristics of STS, as opposed to the taylorist-fordist paradigm.

Table 1. Brief characterisation of STS as a new paradigm and as a changed personal attitude, according to Eijnateen (1993).

Old Paradigm	... Give up
Redundancy of parts External coordination and control Autocracy Fragmented socio-technical system Technological imperative – man as extension of machine Organisational design based on total specification Maximum task breakdown, narrow skills One person-one task Alienation	Feeling of having learned it all Reductionist thinking Dependence on procedures False simplicity It is 'they' who are to blame Virtue of being certain Belief in stability
New Paradigm	**... New Reality**
Redundancy of functions Internal coordination and control Democracy Joint optimisation of the socio-technical system Man is complementary to machine, and a resource to be developed Optimum task grouping, multiple broad skills Self-managing social system Involvement and commitment	Learning never stops Systems thinking Focus on results Complexity Personal accountability Doubt Continuous change

Beyond STS theory, in the late decades of the 20th century diverse authors tried to identify some principles of 'new production rationales' that shared the principle of autonomy to workers. For instance, Terssac & Dubois (1992) summarized the principles of 'new production rationales': (a) the new rationale is not exhaustive. It no more aims at prescribing all production phenomena, but it intends to facilitate the management of non-planned territories, of aleatory events and of uncertain and variable environment. The new rationale aims at managing informal elements, not to suppress them; (b) men are seen as competent actors, capable of taking initiatives and decision within production systems. Competence and autonomy are no longer residual consequences of automation of production processes, due to the nature of automation as discussed before, but they are principles of organization. Control is needed to guarantee that competence will not generate non pertinent improvised acts. (c) the new rationale also brings flexibility to production systems and work organization. Flexibility is considered a property of modern production systems, due to increased competition based on variety, speed and innovation. In this sense, Veltz & Zarifian (1992) point that in order to be flexible, organization must be capable of learning how to manage unpredictable incidents; it must allow short cycles of decision making; it must favor horizontal coordination and communication; and it must encourage the development of competences.

In the next section we will investigate in more detail three practical aspects of the NFWO that were raised by this new rationale: autonomy; competences; and the adoption of KPI as a control mechanism.

KEY ASPECTS OF NFWO: AUTONOMY, COMPETENCES AND KEY PERFORMANCE INDICATORS

Even if more radical new forms of work organization such as STS has a limited diffusion in production systems, perhaps due to its questioning of the division of power established by the classical forms of work organization, the principles of NFWO may be found widespread in manufacturing and service environments, as well as in administrative functions.

Thus, the need for rapid answers in work situations or processes whose demands are permeated by variability, which includes situations such as high contact service (in which the customer is one source of variability) or innovation activities, leads organizations to grant greater autonomy to workers, or, at least, to extend the worker's decision-making field, even if within working conditions and/or organizational spaces predetermined by other actors. There would be greater 'discretion' to the workers in this condition (Maggi, 2006). In this sense, the result of work depends on the engagement of workers – as Zarifian (1999) affirms, in this situation the work goes back to the worker; it is impossible to predict a task independent of who will perform it, as pretended in Taylorism.

As a result, the increase in workers' autonomy comes along with a concern with the qualification or competence of workers, so that both the daily, regular work and the unexpected activities can be treated in the best possible way. In a way, competence replaces old standards and procedures.

Strictly speaking, the concept of competence is not equivalent to the formal qualification; qualification is part of the competence. In this regard, one can identify at least two major theoretical frameworks that have distinct concepts on competence: one sees competences as attributes possessed by employees (such as skills, formal qualifications and attitudes), regardless of their work situation (for example, Boyatzis, 1982). Other framework states that competence exists only as practice, being defined as a 'responsible act' (Le Boterf, 1995), a practical understanding of work situations with which the worker is faced and on which occurs the initiative and responsible action of the worker, eventually engaging coworkers in action (Zarifian, 1999). This practical understanding can be supported by explicit knowledge previously acquired, but such knowledge will be transformed by the very competence in action. This theoretical framework, which we prefer, treats the competence not only as an individual issue, but also as an organizational issue: the competence only exists when the individual acts in the organization. Put in another way, if the organization interdicts initiative and action, competence does not appear. So, when we adopt such a concept of competence, we must abandon the Taylorist concept of 'selection of competent workers'; selection is not sufficient. It is necessary to allow the continuously building of competence, in the practice of work.

On the other hand, the management control in such circumstances is not carried out primarily through the comparison of actual work to standard operating procedures, but rather by the results achieved that are measured by performance indicators. The counterpart of the expansion of the scope of 'responsible action' of the worker is the control by key performance indicators (KPI), which makes 'responsibility' a concrete issue. In fact, KPI systems, such as the Balanced Scorecard (Kaplan and Norton, 2001a; 2001b), proliferate among contemporary organizations, both in private companies and public companies. Particularly in the public sector, the management by KPIs has become synonymous with 'good management', with the 'new public management' best practices that would be characterized by reducing bureaucracy with managerial control (Paula, 2005; Bresser-Pereira, 2000).

In KPI systems, a set of indicators is defined, on the basis of the strategic objectives of the company, and deployed throughout the various hierarchical levels and the different functions. Indicators will operate as coordination mechanisms (Mintzberg, 1983), communicating to employees what is expected of them, directing their actions, assisting their decision-making and motivating them to improve besides, of course, controlling them (Kaplan & Norton, 2001a; 2001b). Every aspect of the company's strategy could be reflected in the KPI systems: costs, quality, flexibility and even innovation; indeed, companies begin to measure aspects such as 'numbers of ideas per worker', or 'number of published papers per engineer' as proxies for innovation. In addition, the system of KPIs is often related to a system of rewards, so that compliance with goals will generate consequences for the remuneration or the workers' career plan.

The notion of a 'system' of indicators has a double meaning: first, the company has different stakeholders whose goals can have multiple natures, and the system should contemplate all of them – in the BSC, for example, indicators should reflect the interests of the shareholders (financial), the interests of customers, the internal needs of the operation and the vision for the future of the company (learning and growth perspective). To Kaplan and Norton, these 'four perspectives' must be balanced in the system of indicators. In addition, being a 'system' means that the indicators of each department or organizational function and of each hierarchical level should be related to each other, so that the goal of a department does not interfere negatively in the goal of other department. For example, achieving the goal of the sales sector should not mean to spoil the goals of production, for instance, to reduce overtime work. Thus, in practice it is a very complex system that presents to the worker a myriad of performance indicators whose goals must be achieved.

To sum up, NFWO rose as opposed to the classical models of organization; in fact, they present assumptions that are quite different to those of Taylorism and Fordism. However, the NFWO can also load paradoxes, particularly in a globalized economic environment in which competition is extreme and the logic of financial capital predominates over productive capital, in a process known as financialization (Zilbovicius & Dias, 2005). The next section will explore some of these paradoxes and their consequences for safety and health.

PARADOXES IN NFWO

KPI Versus Autonomy

A first issue concerns the issue of autonomy, which is an important assumption and practice in NFWO, versus the need for control of results to shareholder brought by increased competition and the predominance of financial capital. The question can be stated in a simple way: investors demand the greatest possible transparency about decisions made in the company, so that they can assess, as continuously as possible, the relevance of keeping their investment or disinvest in the company. So, it is up to the company to define the ways of ensuring such transparency. One of the ways to ensure transparency in operating levels would be relying on standardization of processes as the main coordinating mechanism, which would amount to an increased bureaucracy, as did Taylor in his time. As we have seen, standardization is reduced in NFWO, although it can still be found in ISO 9000 certifications and the like. However, in the present competitive environment, standardization generates dysfunctions that are extremely counterproductive from the point of view of the shareholder remuneration. That is why other forms of control can be introduced, weakening the standardization as the main control instrument; the performance indicators can occupy such space, even though the standardization may not completely disappear.

In order to control through performance indicators and ensure greater transparency, information and communication technologies (ICT) contribute in an almost fundamental way. New ICT allow remote and real time control over workers (Zuboff, 1994). For instance, we recently investigated the case of a water service public company, traded on the São Paulo Stock Exchange, which has introduced an electronic system of reading service consumption and instant issuance of invoices to customers. The main intention of introducing such a system was to reduce working capital as simultaneously reading and issuing the invoice reduced drastically the time between reading the consumption and the payment by the client (from 21 days in small towns and 12 days in big cities for 9 days in both cases). At the same time, the electronic reading system facilitated the control over the workers, who carry out their activities remotely (in the consumption places: residential and commercial buildings), once the system records the time of each reading, the worker displacement interval between residences and intervals for lunch, for instance.

This control system is utilized in all companies inserted in global supplier networks in the form of contract manufacturing. In this case, KPI systems, notably concerning costs, quality and production rates, and ISO certification are the basic conditions for the perfect interchangeability of components made in all countries, usually in peripheral ones. Despite all circumstances and differences, the quality, costs and production rates must be equivalents, due to competition pressures. In this network structured in various levels, the top industries are, not rarely, supported by the work in subhuman conditions. This reality is well known, but this necessary relation with the corporate organization and governance is not sufficiently stressed. The KPIs are now extended to environmental and social criteria, trying to grasp these and others 'unfair' competitive practices, but not only the KPI system is insensible to the real condition of the underground work, but also it is its principal determinant. In the modernity, we always can find workers in slavery like conditions not only in the agricultural activities but in the main industrial cities.

In this way, a major contradiction is present: in conditions of fierce, global competition, combined with financialization, the control and transparency requirements are placed. Thus, is it possible that real autonomy exists? In fact, the discussion is not new. Deleuze (1990) advocated the passage of a 'disciplinary society', described by Foucault (1975), towards a 'society of control', in which the discipline of the bodies retrieved from the confinement and by the discipline of timetables, movements and procedures shall be replaced by 'continuous', malleable control, provided, in particular, by the new ICTs and based on numbers, indices and every data associated with each object (or subject) to be controlled. Zarifian (2003) describes how this control is reflected in working conditions by means of flexible working time; malleable space; and malleable subjective engagement.

So, although there is a need to exercise a certain degree of autonomy, in order to meet the criteria of competition and capital growth, at the same time worker is controlled through a multitude of performance indicators. The space of action that is supposed to be large can, in fact, continue to be quite restricted.

This is reinforcing by the extension of the indicators to safety and health management systems. Here, we are in the worst of the worlds: performance control by indicators and by safety standards. The obliged transparency and publicity of the results, including social, environmental and safety/health indicators creates an 'opacity' of the facts in the bottom of the production chain (in the 'shop floor'). The KPI system is coupled with reward system and bonuses in a manner that the bottom events influence the goals of the executives in the top. So, by an opportunistic (and perverse) adaptation and group coercion in all levels, including the victim itself, the accidents and occupational diseases are masked to preserve the appearance of good performance and save the bonuses.

Managers, and some authors, are incapable to perceive that this opportunistic behavior is not an immoral one, but a necessary consequence of the governance rationale founded in KPI systems. If frequency and gravity indicators of health and safety are messed with production and financial performance, then sub-notification by workers and supervisors in the shop floor is rational. This behavior is reinforced by the formalist method of accident analysis. Due to the unavoidable distance between tasks' prescription and real work, we can always find a deviation behavior in the causal chain of an accident, because workers are engaged in the solution of a problem that cannot be solved solely with the actual procedures. Frequently, the problem is raised by norms conflicts, called for the discretionary competence of the workers. Eventually, these decisions may be involved in a sequence of events that precede an accident and the worker is invariably judged as negligent or imprudent. So, without an objective reason to justify his choice, the victim prefers, if possible, to hide the accident. The obligation to publicize the health and safety indicators' implies the production of the good indicators as a parallel activity, instituting a new form of invisibility of the real work conditions.

Competence Development Versus the Need for Short-Term Results

A second problem is related to the process of competence development. As discussed earlier, NFWO encourage the development of workers' competences, and allow for the exercise of such competences in relatively autonomous spaces of action. However, we must emphasize that competence building – be it an individual, group or organizational competence – is necessarily a long-term process. Even if one adopts a point of view (which we do not share) according to which it would be possible to 'buy' competences on labor market, selecting staff according to the competences that the company craves, still it would be necessary a time for individual competences to fuse within the organization.

However, in a situation of fierce competition and the existence of the imperative to generate shareholder value, the short-term result usually predominates, given that one of the key metrics of performance is the value of the action, monitored on a daily basis and the rewards are distributed annually. Indicators of 'value creation' that point a bad performance can induce the shareholders the crumble of their investments more quickly, which can lead to the closure of the company. This kind of 'short-term syndrome' is described, for example, in Ezzamel *et al.* (2008), and is also discussed (and criticized) by Veltz (2000), Kaplan and Norton (2001a), Gaulejac (2005) and Sennett (1998).

We argue that the adoption of models based on competence, such as the proposed by Zarifian (1999), does not hold up combined with pressures for short-term financial results. Nevertheless, if the competitive environment demand differentiation strategies, the competence of workers can be required. Then, another paradox comes: the worker must develop his or her competences, and demonstrate them, in the short term, which is impossible. If 'value' is not generated (according to the metrics adopted), this can be interpreted, in the context, as an issue related to workers, that would not be competent – even though some metrics, like the share price, may have nothing to do with internal actions of the company, since there may not be a cause-and-effect relationship between managerial decisions and the share price (which is held also in the stock market). Also if the goals presented by the KPIs are not achieved, this might be seen by the management as a lack of competence of the worker. It is up to the employee to deal with the problem and solve it, thinking of his or her 'employability'. In this sense, Clot & Zarifian (2009) draw attention to the fact that management by indicators may make workers' activity (i.e. real work) invisible; in their words, real work becomes a 'blind point' as "between goals on one side, (and) results on the other side, we organize the disappearance of the essential: the work itself. Workers must achieve results under the tension of numbers, and the recognition of their efforts on work disappears" (Clot & Zarifian, 2009 – translated by the authors). In other words, management may not recognize that workers may be competent even if results are not achieved – workers might have strived to achieve the goals, but something else might have happened in the way that frustrated their attempts.

Engagement in a Flexible World

Another point concerns the question of engagement. From what was exposed about the NFWOs, it may be assumed that their adoption produces and at the same time assumes greater engagement on the part of the worker; indeed, as described in Table **1**, 'personal accountability' and 'involvement and commitment' are keywords in NFWO such as sociotechnical systems. At Google, a contemporary paradigm of innovative firm, workers are treated 'entrepreneurs' who must come up with new ideas (Savoia & Copeland, 2011). Also, Zarifian's (1999) definition of 'competence' holds the worker's initiative when facing work situations, which can occur only through his or her engagement; the worker shall decide, in this context, when to engage, without needing a superior order for this to occur (Zarifian, 2003). The author draws attention to the fact that such engagement is possible only if there is a reciprocal 'engagement' on the part of the company in relation to the employee *i.e.* a company's commitment to its employees (Zarifian, 1999).

To implement an effective prevention program this reciprocal engagement is also crucial. If the workers are prejudged as guilty, the accidents analysis remains superficial and the real reasons of deviations are not revealed and prevention cannot progress (Terssac & Mignard, 2011). The more distance grows between management and the workers in the shop floor, the more it is difficult to accept the justifications expressed by workers. Due the impossibility to assert the objective causes of an action, the shared experience is necessary to accept subjective explications as reasonable ones. This is an affair engaged between workers and first level hierarchy that is mined by the moralistic judgment of the superior levels. In this way, we create a sort of social mechanism that make impossible the effective manifestation of the responsibility and autonomy of the workers, like a 'systemic thought' (Beck, 2008) that make individuals powerless to face the great risks in the modern production. Moral judgments forbid real ethical commitments and comprehensive dialogs within a hierarchical organization. Thus engagement is necessary to contemporary production not only from the point of view of economic performance but also from the point of view of health and safety aspects.

At the same time, one of the goals of contemporary company is flexibility. In addition to the flexibility of production processes and products, flexibility concerns also labor: workers must be flexible in carrying out functions; the working time is flexible; but also the quantity of workers is flexible. Indeed, in a company's capital structure, the 'human capital' is not a fixed asset and rarely, in the case of direct labor, is considered a specific asset. Not by chance, it is commonplace to note that, during crisis situations, one of the first corporate attitudes is to reduce the number of workers. Still, it is common that the announcement of major layoffs lead to high in the share price of the company concerned (Gaulejac, 2005; Plihon, 2003). How could one expect workers' engagement in a similar context? There is thus a new paradox: engaging even if the constant threat of dismissal hangs around. Once again, it is up to the worker to take care of his or her 'employability' as the only possible means of escaping such threat. Managing this risk becomes part of everyday work (Amoore, 2004).

All these constraints and organizational pressures submit all the workers (and here we need also consider the managers as workers) to perform the best despite the limited resources. Because of these excessive exigencies, bullying at work became an overall phenomenon in modern organizations. Several authors discussed this abusive work relationship and its effects on workers' health. However, the main approach explains bullying at work essentially as an event of a moral and psychological nature (for a divergent interpretation see Vieira *et al.*, 2012). In fact, bullying is no more than the top of an iceberg that occult these systemic mechanisms that creates the perversity of the current management patterns. Insofar as this mechanism is presented as moral behavior, we remain incapable to explain the objective causes of pathogenic features in modern organizations. The perversity in the management and between workers is merely the consequence of structural forces in the organization that link the individuals in a performance based organization, supposedly consensual and coherent but, in fact, crossed by paradoxes and conflicts that block the collaborative work.

Gaulejac (2005) develops a contemporary and critical analysis of the new management models that establish impossible goals and succeed, in spite of this, mobilizing workers, even though they are not given the appropriate working conditions for improve performance. Goals ever increasing, contradictory goals, paradoxical injunctions, guilty attribution and individualization of results constitute the backbone of work in the 'hypermodern organizations' (Gaulejac, 2005). Under these pervasive phenomena, the author identifies a "subjectivist conception of action, the ideology of self-realization, which transforms the social contradictions in relational problems" (Gaulejac, 2005, translated by the authors).

The moralization of the bullying behaviors' takes away the problem of the organization and also the impossibility to confront it. Bullying becomes a juridical problem to repair the moral damages, without consequences to the organization and management models. The goals continuously growing are not issues of the will of evil executives and administrative boards, but they only express the insatiable nature of the economic value in valorization process – that is intensified by the processes of globalization and financialization of production.

> "In this sense, the growing production of material wealth exacerbates the problem because the accumulated wealth becomes an assumption of a new cycle of economic growth, a starting point that requires further productivity increases and so on indefinitely. The contradiction that the objectified labor, the immense material wealth accumulated, can only reproduce in an amplified way absorbing living labor, which becomes the ever closer basis of new cycles of development, after several mediations, manifests itself in the pressure on workers to continuously improve their performance" (Vieira *et al.*, 2011).

IMPACTS OF AUTOMATION AND WORK ORGANIZATION ON THE SUBJECTIVITY AND WORKER INVOLVEMENT

In this section we will discuss how the new productive rationality presented in the introduction impacts on work organization and on workers' subjectivity in a particular environment: the automated production processes. These impacts depend on how one understands the need for responsibility and personal involvement in decision-making; on how one deals with the possibility of error; and on how global time is divided between time of action and time for reflection on the action. The need for a worker with a different profile to meet the demands of new productive rationality can now be based on objective needs made by the way the automated production system works and how it should be managed. To understand singular and unexpected events, out of routine operational standards, calls for a new form of communicative and reflexive rationality (Zarifian, 1985).

This issue was discussed by Zarifian (1990; 1993; 1995) who proposes the concept of 'qualifying organization' (*organisation qualifiante*) to give an account of the specific needs of the new productive rationality of automated production systems. One of the highlighted aspects concerns the forms of calls and interactions of different knowledge and experiences of the agents of production, which would meet within working teams with different goals, either operational or project goals (quality improvement, optimization; diagnosis of panes in TPM-Total Productive Maintenance etc). In order for these new forms of management and organizational models to generate results, it would be necessary to provide the management of a new rationality of communicative nature rather than instrumental, which, however, is faced with various obstacles. To give just one example, the diagnosis of a breakdown in a complex, automated system, when it involves circumstances fleeing to already formalized knowledge, presupposes the articulation of several informal knowledge (production, maintenance, quality, purchasing), whose conditions of possibility (cooperative work, personal involvement, dialogue, mutual understanding) are not yet found in companies, subject to the capitalist mechanisms of control in which prevails the hierarchical organization of work. There is no communicative rationality in a situation where one of the parties is from the beginning placed in situation of social subordination, *i.e*, submitted to the power of others. This is one of the main causes of distress of operators of continuous processes.

We know that the work in continuous processes industries is potentially pathogenic, due to the own characteristics to the process industry, in particular the risks of explosions and exposure to chemical agents, whose health effects are still unknown. These are perhaps the most obvious aspects, revealed by surveys conducted in nuclear power stations and chemical industries, which add the journalistic reports of accidents of catastrophic dimensions (Bhopal, Seveso, Three Mile Island, Chernobyl, among others.). Less well known is the everyday wear and tear caused by temporal requirements related to the uninterrupted flow and the uncertainties arising from the complexity of the facilities. This form of wear (commonly attributed to stress) hardly manifests as occupational specific, already recognized pathologies, but first of all as psychic suffering. In what follows we analyze two striking aspects of this suffering related to the everyday work of production flow control: the experience of time, given the relative uncertainty of events, and responsibility for decisions to be taken by operators, given the inadequacy of the rules laid down.

One of the striking features of the new automated systems is the uneven distribution of workload concerning moments and situations, what Zarifian (1995) calls possible situations (events) that occur in a not predictable manner. These moments are characterized not only by the emergency situations (e.g. severe crashes, stops), but also for situations where several small problems or even routine situations occur simultaneously (Daniellou & Boel, 1983; Daniellou, 1986). Given the oscillating character of workload and system requests, there will always be, at certain times, the other side of the coin: under load situations, in which operators are being less requested in their psycho-sensorial functions. In these moments, operators, contrary to what can imagine the hierarchy or a rushed visitor, are also working, actively monitoring certain parameters and attentive to alarms that may soar, indicating an unexpected event. Therefore it is not a useful, free time, during which the worker could devote to other activities, particularly those that require reflection; for example, to think about the causes of a recent incident or how to react to it.

It is important to retain this suggestion concerning the subjective experience of the operators in the face of uncertainty and the possibility of committing errors. As the uncertainty is, ultimately, an ontological given impossible to be excluded from production, to eliminate this source of suffering it is necessary to create conditions to deal with unforeseen events and results, which demands, in particular, interventions within the framework of work organization and management. Given the difference between knowledge and representations of engineers and workers, one of the important questions is whether the distance between them can be eliminated within a hierarchical organization of work, that is, with a differentiated allocation of tasks, responsibility and knowledge; or, if not, what conditions are necessary to establish a real dialogue between actors of production and forms of knowledge that each one of them hold. In summary, the uneven distribution of free time to reflect on the process is one of the main obstacles to the establishment of a new productive rationality.

As the competence and autonomy of workers develop largely in hiding, only their errors gain visibility, that is, when they cannot control the process after taking a decision contrary to the guidance they have received. When nothing extraordinary happens, it is as if the operation had been assured by strict obedience to the guidelines of his superiors, which are seen as responsible for the proper functioning of the process. Thus a cleavage is operated in the personality of the operators, who become responsible only for errors, not being recognized by the right decisions. In these situations, workers live together with a radical ambiguity that exists in any hierarchical relationship: they do not like to work with their chiefs giving guesses but, when they are alone at night, feel the lack of his superiors in moments when they must take an important decision, as to stop a production unit.

NEW FORMS OF INDUSTRIAL ORGANIZATION: GLOBAL NETWORKS AND VALUE CHAINS

In addition to changes in the work organization, recent cross-border organizational transformations have caused impacts for labor. Strictly speaking, such transformations change the very concept of organizational boundary. Among the main transformations, we will discuss the networks or supply chains forms of organization and the global relationships between companies.

The combination of high capital costs; high technological complexity; shorter product life cycles and financial and productive globalization fostered, notably from the end of the 20th century, the vertical disintegration of organizations, generating productive configurations in local or global networks or chains. High capital costs make it more expensive to maintain a large corporate structure, leading companies to outsourcing of no-core activities (Prahalad & Hamel, 1990). In many cases, outsourcing is performed towards different countries, generating global production networks (GPNs), a typical configuration of multinational companies (MNCs) (Ernst & Kim, 2002; Gereffi *et al.*, 2005). At the same time, the rapid technological development, toward more complex technologies, makes it necessary for companies to specialize in a few technologies, whose evolution can be perfectly mastered (Veltz, 2000; Prahalad & Hamel, 1990). Also, fierce competition, including at the global level, suggests that new fields of value in products and services are due to the combination of these technologies, requiring the experts companies, somehow, to create long-lasting relationships in order to integrate their operations (Veltz, 2000).

Conceptually, networks would mean a set of companies that relate without strong hierarchy between them. On the contrary, networks encourage cooperation and solidarity, a sense of generalized reciprocity that is common when actors share some background (be it geographical, ideological, ethnic or professional). The presence of trust decreases the need for monitoring, control, hierarchy (Powell, 1990).

Production chains would feature the existence of hierarchy, that is, some of the actors of the chain drive it. For example, there would be production-driven chains and, on the other hand, chains driven by companies that market the final product (Gereffi, 1994). More recently an alternative classification was proposed by the same author (Gereffi *et al.*, 2005): chains could be modular value chains, in which suppliers produce according to more or less detailed specifications of customers, but dominate their process technology; relational value chains, in which there are complex interactions between buyers and sellers, creating mutual dependence;

and captive value chains, in which small suppliers are dependent from larger buyers. In this situation is it not uncommon that these large buyers monitor and control the sellers' processes, including work organization.

The internationalization of productive chains and networks at the end of the 20th century raised the question of how the activities of production systems would be distributed among the countries. If the distribution of global production networks could mean new labor opportunities for emerging countries, the question is: which type of labor are these opportunities made of? The international division of labor in these networks and chains is thus an important discussion, both from the point of view of the central countries and from the point of view of emerging countries. At first, such a division seemed to point to the centralization of activities considered more 'noble', from the point of view of adding value, in developed countries, while activities that add less value, more labor intensive ones, would be shifted to developing or peripheral countries.

One central characteristic of capitalist economy is the unequal development and wealth appropriation between social classes and countries. Such inequality is set up in an international division of labor, with different attributions and wealth distribution – and also with an unequal distribution of risks. The hazards exportations to peripherals countries are a usual strategy of multinational firms (Castleman, 1979). The delocalization of entirely plants is the practice in chemical and metallurgical industries due the environmental costs. With increasing labor costs, the manufacturing industries, labor intensives, also migrate to developing countries. Without social control capable to oppose the capital tendencies to increase their profitability, the human and ecological damages were intensified. Then, accidents and occupational diseases are redistributed in a world scale. In specifics production chains, the onus is all located in developing countries and the bonuses in the central ones.

One paradigmatic case in the recycling chain, where an important social and economic activity is supported by waste pickers in an 'informal' work, in subhuman conditions, that provides materials to high technological industries, like aluminum, chemical and electronics. Not enough, the incineration industry begin to dispute the attractive waste market in the BRICs (Brazil, Russia, India, China and South Africa). In Brazil, various municipalities are planning to install incinerators, presented by Europeans companies that find social resistances to implement this technology in home because the ecological and public health risks.

In general, Brazil occupied a place in the international division of labor that exposes its workers to greater risks. Another example is the meat industry. The meat industry is considered a success case in internationalization of Brazilian firms, supported by public funds. Nonetheless, this 'successful' economic sector is marked by the one of the higher rates of the work related musculoskeletal disorders (WMSDs). Recently (April 2013), the Brazilian Labor Ministry proclaimed a specific regulation (NR-36) to the slaughterhouse work conditions. This official standard determines pauses and others limitations to work intensification, but leaves untouched the essential of work organization that is like a Ford's line. We can forecast the ineffectiveness of the measures to prevent musculoskeletal diseases. To promote effective prevention, the 'disassembly line' could be broken into cellules of production without paced rhythm by the conveyor and longer work cycles, according the principles to sociotechnical work organization that we presented previously. But to do this implies retrocede the control over work time. Contrary to the pretension of absolute control on time and movements, the sociotechnical organization creates obscurity in work realization by the principle of minimal specification. This is one of the reasons to restrict the diffusion of this NFWO in this 'successful' industry.

In the early 20th century, however, the growth of emerging countries, in particular the so-called BRICS, and implementation of public policies for technological catching up on the part of these countries, brought greater complexity to the international division of labor panorama. Indeed, the international division of labor seems to depend on conditions relating to industrial sectors, countries and even the strategies of companies (Dias *et al.*, 2012; Dias & Salerno, 2004). So, there are many possible configurations for the global chains and networks, none of them relentless. If the export of risk activities to developing countries is a reality, it can be circumvented by means of appropriate public policies.

We can see the possibilities of the public regulation in the case of outsourcing. It is usual the statement that outsourcing is cause of accidents and occupational diseases. Nevertheless, "the precise mechanisms associating outsourcing and occupational health are not well known yet. Even being the relations between outsourcing and health apparently evident, it is not easy to demonstrate the existence of direct and specific causal relations between the change of the contractual relations and occupational diseases. The various aspects of juridical, social and economic deregulations, although real, do not explain themselves the increasing damages on the external workers' health" (Santos *et al.*, 2009) if these damages were also observed before these organizational change. This was demonstrated in the case of privatization of the urban waste services in a Brazilian city. The deregulation of the work relations that came along with outsourcing of urban cleaning services produced a growth in the occupational diseases and accidents of the street sweepers. These specifics health damages occur because the deregulation of contractual relations leave to a deregulation of the work activity itself. In particular, it was shown that the collective strategies were perturbed by the supervisors' arbitrariness of the private enterprise to recompose the teams. The collective know-how of the groups of street sweepers working together for years was broken when one of them was reallocated because disciplinary punishment. With this group reallocation, the sweepers were also relocated in other urban sector, where they have no specific practical experience necessary to prevent accidents and to do the work more efficiently, so without overloading themselves and the teams' companions. This experience is so important that the urban sector was subjectively appropriated and called by the sweepers as '*my strecht*' (*meu trecho*). This appropriation is more than a symbolic attribution, but essential to preserve physical and mental health. The group reallocation and the consequent health problems that followed could have been avoided if regulations or other public instruments had been associated with the privatization process, within the context of a particular public policy.

FINAL REMARKS: THE OTHER SIDE OF WORK ORGANIZATION

From his beginning, capitalism is simultaneously production of wealth and poverty, unequally distributed between individuals, classes and countries. In the modern 'risk society' (Beck, 2008), in a great measure, all the individuals, independently of the class, is submitted to risks that encompasses the society at all. Nuclear energy, genetically modified organisms, viruses' epidemics, catastrophic accidents in chemical and transport industries affects the individuals without concerns to social status and properties. In this sense, the management system is a democratic plague that affects managers and workers. The mental suffering and psychological diseases are equally distributed among all members of the hypermodern organization: the suicide affects both the unemployed workers and the engineers stressed by goals impossible to reach. This fact only emphasizes the systemic nature of the causal determination of the health damage in NFWO. In despite of the fact that the rewards are much more attractive to executives than to workers, they all are submitted to the same alienated work; they are not able to master their acting according to their goals. The contrary is what happens: the fixed goals determine their actions.

These organizational changes carry out a change in the standard profile of occupational diseases. The damages become more unspecific, with a symptomology similar to common diseases. In consequence, is more difficult to establish nexus with work conditions, like traditional occupational diseases with unique or multifactorial, but identifiable, causes. Now, everything happens as if the immaterial work changes creates immaterial disease factors and, also, immaterial damages, where prevail the mental and psychological ones. To determine a precise nexus between a disease and an occupational disease is always a controversial issue. Nonetheless, in this case the difficulty is potentiated by the subjectivity of the symptoms and, further, by the general profile of these diseases. Cardiovascular disturbs, depressions, phobias, sleep disturbs, psychosomatics diseases, cancers and immunologic dysfunctions are phenomena so linked to general forms of life that is difficult to differentiate the part of work conditions determinations and the general stress of the urban life conditions. Perhaps, this presumption that work and private life are isolated spheres of the society should be leaved out if we pretend explain the actual interrelationships between work and health. Furthermore, one feature of the recent transformations of work organization is the confusion of the frontiers between worktime and space and the familiar and personal spaces and time. Supported by ICT, the homework is growing and the rest time becomes waiting time disposable to satisfy the needs of flexible organization.

To discuss work organization is always delicate due the multiplicity of factors implied. Nonetheless, in our discussion and in the cases presented here some common orientations emerge:

- The conflicts between control, hierarchical organization and autonomy;

- The paradoxes arising in the subjective workers' engagement without reciprocity of the organization;

- The perverse effects of the '*gap of the procedures*', that opens space to the hierarchy arbitrariness that may take the form of KPIs;

- The obstacles and limits to develop an effective communicative rationality, nonetheless solicited by the new forms of work organizations.

Here we want to stress another point: the obscurity always carried with work activity. In fact, all the forms of work organization contain a '*dark side*', but each has and produces his specific spaces of the invisibility. These different forms of invisibility are one of the most important determinations of the health damages in work situations. They are source of the excessive workloads and, inversely, source of autonomy and workload regulations. In the sociotechnical work organizations, this obscurity is intentionally produced, although in some defined limits. If the goals are consensually defined, this space of autonomy is source of self-realization and health; if the deregulation is a pretext to impose unrealizable goals, it is source of stress and physical and psychological damages. It is the organizational power relations that give the direction. For this reason, the actual limits imposed to the workers' autonomy are so difficult to enlarge.

ACKNOWLEDGEMENTS

None declared.

CONFLICT OF INTEREST

The authors confirm that this chapter content has no conflict of interest.

REFERENCES

Amoore, L. (2004). Risk, reward and discipline at work. *Economy and Society, 33*(2), 174-196.

Beck, U. (2008). *La société du risque: sur la voie d'une autre modernite.* Paris: Flammarion.

Boyatzis, R. E. (1982). *The competent manager: a model for effective performance.* New York: Wiley.

Bresser-Pereira, L. C. (2000, julho). A reforma gerencial do estado de 1995. *Revista de AdministraçãoPública, 34*(4), 55-72. Retrieved from http://www.bresserpereira.org.br/papers/2000/608-RefGerencial_1995-RAP.pdf.

Castleman, B. I. (1979). The export of hazardous factories to developing nations. *International Journal of Health Services, 9*(4), 569-606.

Cherns, A. (1987). Principles of sociotechnical design revisited. *Human Relations, 40*(3),153-162.

Clot, Y., & Zarifian, P. (2009, 18 décembre). Evaluation des performances, point aveugle. *Le Monde.* Retrieved from http://www.lemonde.fr/idees/article/2009/12/18/evaluation-des-performances-point-aveugle-par-yves-clot-et-philippe-zarifian_1282672_3232.html

Coriat, B. (2000). The 'abominable Ohno production system': Competences, monitoring, and routines in Japanese production systems. In G. Dosi, R. R. Nelson, & S. G. Winter (Eds.). *The nature and dynamics of organizational capabilities* (pp. 213-243). New York: Oxford University Press.

Daniellou, F. (1986). *L'opérateur, la vanne et l'ecran.* Montrouge: ANACT.

Daniellou, F. (2005, September). The French-speaking ergonomists' approach to work activity: Cross-influences of field intervention and conceptual models. *Theorethical Issues in Ergonomics Science, 6*(5), 409-427.

Daniellou, F., & Boel, M. (1983). *L'activité des opérateurs de conduite dans une salle de controle de processus automatisé.* Paris: CNAM.

Daniellou, F., & Rabardel, P. (2005, September). Activity-oriented approaches to ergonomics: Some traditions and communities. *Theorethical Issues in Ergonomics Science, 6*(5), 353-357.

Deleuze, G. (1990). *Pourparlers.* Paris: Éditions de Minuit.

Dias, A. V. C., Pereira, M. C., & Britto, G. (2012). Building capabilities through global innovation networks: Case studies from the Brazilian automotive industry. *Innovation and Development, 2*(2), 248-264. Retrieved from http://web.cedeplar.ufmg.br/cedeplar/seminarios/ecn/ecn-mineira/2012/arquivos/Building%20capabilities%20through%20global%20innovation%20networks.pdf

Dias, A. V. C., & Salerno, M. (2004). International division of labour in product development activities: Towards a selective decentralisation?. *International Journal of Automotive Technology and Management, 4*(2-3), 223-239.

Eijnatten, F. M. (1993). *The paradigm that changed the workplace.* Stockholm: The Swedish Center for Working Life.

Ernst, D., & Kim, L. (2002). Global production networks, knowledge diffusion, and local capability formation. *Research Policy, 31,* 1417-1429. Retrieved from https://www.eastwestcenter.org/fileadmin/stored/misc/Global_production.pdf

Ezzamel, M., Willmott, H., & Worthington, F. (2008, February/April). Manufacturing shareholder value: The role of accounting in organizational transformation. *Accounting, Organizations and Society, 33*(2-3), 107-140.

Foucault, M. (1975). *Surveiller et punir*. Paris: Gallimard.

Gadrey, J., & Zarifian, P. (2002). *L'émergence d'un modèle du service: Enjeux et réalités*. Rueil-Malmaison: Éditions Liaisons.

Gaulejac, V. (2005*). La societé malade de la gestión*. Paris: Seuil.

Gereffi, G. (1994). The organization of buyer-driven global commodity chains: How U.S. retailers shape overseas production networks. In G. Gereffi, & M. Korzeniewicz (Eds.). *Commodity chains and global capitalism* (pp. 95-122). Praeger: Westport.

Gereffi, G., Humphrey, J., & Sturgeon, T. (2005, February). The governance of global value chains. *Review of International Political Economy, 12*(1), 78-104. Retrieved from https://rrojasdatabank.info/sturgeon2005.pdf

Gorz, A. (2005). *O imaterial*. São Paulo: Annablume.

Guérin, F., Laville, A., Daniellou, F., Duraffourg, J., & Kerguelen, A. (2006).*Understanding and transforming work: The practice of ergonomics*. Paris: ANACT.

Haspeslagh, P., Noda, T., & Boulos, F. (2000). *Are you (really) managing for value?*. Fontainebleau, France: INSEAD.

Kaplan, R. S., & Norton, D. P. (2001a, June). Transforming the Balanced Scorecard from performance measurement to strategic management: Part II. *Accounting Horizons, 15*(2), 147-160.

Kaplan, R. S., & Norton, D. P. (2001b, March). Transforming the Balanced Scorecard from performance measurement to strategic management: Part I. *Accounting Horizons, 15*(1), 87-104.

Keiser, V. (1991). Work analysis in French language ergonomics: origin and current research trends. *Ergonomics, 34*(6), 653–669.

Le Boterf, G. (1995). *De la compétence: Essai sur un attracteur étrange*. Paris: Éditions d'Organisation.

Maggi, B. (2006). *Do agir organizacional*. São Paulo: Edgard Blücher.

Mintzberg, H. (1983). *Structure in fives: Designing effective organization*. Englewood Cliffs, New Jersey: Prentice Hall.

Paula, A. P. P. (2005). *Por uma nova gestão pública: Limites e potencialidades da experiência contemporânea*. Rio de Janeiro: FGV.

Plihon, D. (2003). *Le nouveau capitalisme*. Paris: La Découverte.

Powell, W. (1990). *Neither market nor hierarchy: Network forms of organisation*. (12). [S.l.]: Research in Organizational Behavior. Retrieved from http://woodypowell.com/wp-content/uploads/2012/03/10_powell_neither.pdf

Prahalad, C. K., & Hamel, G. (1990, May/June). The core competence of the corporation. *Harvard Business Review, 68*(3), 79-91. Retrieved from http://www1.ximb.ac.in/users/fac/Amar/AmarNayak.nsf/dd5cab6801f1723585256474005327c8/456e5a8383adcf07652576a0004d9ba5/$FILE/CoreCompetence.pdf

Santos, M. C. O., Lima, F. P. A., Murta, E. P., & Motta, G. M. V. (2009). Desregulamentação do trabalho e desregulação da atividade: o caso da terceirização da limpeza urbana e o trabalho dos garis. *Produção, 19*(1), 202-213. Retrieved from http://www.scielo.br/pdf/prod/v19n1/13.pdf

Savoia, A., & Copeland, P. (2011, April). Entrepreneurial innovation at Google. *Computer, 44*(4), 56-61. Retrieved from http://patrickcopeland.org/papers/EntrepreneurialInnovationGoogle.pdf

Sennett, R. (1998). *The corrosion of character: The transformation of work in modern capitalism*, [S.l.]: Norton.

Terssac, G., Dubois, P. (1992). Les rationalisations: quels choix pour quelles conséquences?. In G. de Terssac, & P. Dubois (Eds.). *Les nouvelles rationalisations de la production* (pp. XVII-XXXIII). Tolouse: Cépaduès-Éditions.

Terssac, G., Mignard, J. (2011). *Les paradoxes de la sécurité: Le cas d'AZF*. Paris: Presse Universitaire de France.

Veltz, P. (2000). *Le nouveau monde industriel*. Paris: Gallimard.

Veltz, P., Zarifian, P. (1992). Modèle systémique et flexibilité. In G. de Terssac, & P. Dubois (Eds.), Les *nouvelles rationalisations de la production* (pp. 43-61). Tolouse: Cépaduès-Éditions.

Vieira, C. E. C., Lima, F. P. A., & Lima, M. E. A. (2012, julho/dezembro). E se o assédio não fosse moral?: Perspectivas de análise de conflitos interpessoais em situações de trabalho. *Revista Brasileira de Saúde Ocupacional, 37*(126), 256-268. Retrieved from http://www.scielo.br/pdf/rbso/v37n126/a07v37n126.pdf

Wisner, A. (1995). Understanding problem building: ergonomic work analysis. *Ergonomics, 38*(3), 595-605.

Zarifian, P. (1990). *La nouvelle productivité*. Paris: L'Harmattan.

Zarifian, P. (1993). *Quels modèles d'organisation pour l'industrie européenne?*. Paris: L'Harmattan.

Zarifian, P. (1995). *Le Travail et l'événemen*. Paris: L'Harmattan.

Zarifian, P. (1999). *Objectif compétence*. Rueil-Malmaison: Éditions Liaisons.

Zarifian, P (2002). L'entreprise de service. In J. Gadrey, & P. Zarifian (Eds.). *L'émergence d'un modèle du service: enjeux et réalités* ((pp. 19-56). Rueil-Malmaison: Éditions Liaisons.

Zarifian, P. (2003). *A quoi sert le travail?*. Paris: La Dispute.

Zilbovicius, M., Dias, A. V. C. (2005). Working for value creation: Some issues on financialisation and new forms of work organization. *Proceedings of* GERPISA International Colloquium, Paris, 13, pp. 1-9.

Zuboff, S. (1994, novembro/dezembro). Automatizar/informatizar: As duas faces da tecnologia inteligente. *Revista de Administração de Empresas,(34)*6, 80-91. Retrieved from http://www.scielo.br/pdf/rae/v34n6/a09v34n6.pdf

CHAPTER 2

Nanotechnology and Risks into the Workers' Universe: Some Critical Reflections

Paulo R. Martins[*] and Richard D. Dulley

Brazilian Research Network in Nanotechnology, Society and Environment (Renanosoma), São Paulo, SP. Brazil

Abstract: This chapter approaches the emerging risks, impacts and challenges posed by scientific and technological development to workers' health and safety, to the environment and to the society, using nanotechnology development as backdrop for the discussion. Nanotechnology encompasses an extraordinary diversity of technological approaches currently under development. It has created a relatively new industry. The goods being produced and introduced into the market are usually innovative and unfamiliar to most consumers. It was chosen as reference to this chapter, due to the fact that nanotechnologies were identified as an important source of known and unknown risks, as an existential risk, and due to the increasing uncertainties posed by these technologies, regarding unique challenges in occupational health and safety, as well as environmental, legal, societal end ethical impacts.

Keywords: Capital contradictions, capitalism, converging technologies, disruptive technologies, nanoethics, nanotechnology, nanotoxicology, new technologies, non-knowledge production, occupational health and safety, precautionary principle, risk acceptability, risk perception, risks, scientific unknowns, societal impacts, uncertainties, unions, workers.

INTRODUCTION

The rapid roll out of new technologies in current days poses a unique challenge to the world of work and to workers themselves, once the lack of information on health and safety impacts make risk management difficult in a traditional way.

*Corresponding author Paulo R. Martins: Renanossoma, Rua dos Pinheiros 1285, ap. 7, CEP 05422-012, São Paulo, SP. Brazil; Tel: +55 11 3812-9954; E-mail: marpaulo@uol.com.br

Marcela G. Ribeiro (Ed)
All rights reserved-© 2014 Bentham Science Publishers

The reflections made in this chapter result from more than ten years of work of the *Brazilian Research Network in Nanotechnology, Society and Environment* (RENANOSSOMA; http://www.nanotecnologiadoavesso.org/) joint to Nanotechnology team project of *Fundacentro* (Fundação Jorge Duprat Figueiredo de Segurança e Medicina do Trabalho; http://www.fundacentro.gov.br), a Brazilian research institute on occupational health and safety (Arcuri, 2009).

It comprises more than two hundred TV programs transmitted over the internet, nine international organized seminars, and several studies published as books, papers and DVD'S. The discussion is mainly focused on issues that should concern not only workers, but the society as a whole.

This chapter aims at reflecting on the emerging risks, impacts and challenges posed by scientific and technological development to workers' health and safety, to the environment and to the society, using the advent of nanotechnology as backdrop for the discussion. Furthermore, it aims at identifying those who benefit from the economic results obtained from the adoption of new technologies that impact on health, safety and environmental working conditions.

EVOLUTION OF TECHNOLOGIES OVER TIME

Currently, the period of time elapsing between one discovery and another more advanced is greatly reduced. The pace of new technologies development has been remarkably intensified from the second half of the twentieth century, until it reaches the current breakneck speed. The result was the emergence of highly sophisticated processes and products, such as minimally invasive surgeries; mobile phones; highly efficient computers; multimedia resources; 3D printers; and nanoengineered products, among others.

In the world's history, by the time that territorial boundaries were mostly well-defined, bearing and expansion of technological boundaries became an important geopolitical strategy. This technological race has played a substantial role on the emergence of contemporary technologies. As a result of its transformation power, new technologies are reordering both human and labor relationships, as well as working environment. And, as any other scientific-driven revolution, it can cause major social disruption, by altering spatial and temporal dynamic of societies (National Science Foundation [NSF], 2001).

The use of electricity, for instance, enables faster communication worldwide. Undoubtedly, electricity is a technology completely inserted into the economic, social and everyday activities of earth inhabitants. Initially, it was seen only as a form of public lighting system. Gradually, however, it has contributed to the development of innumerous other inventions such as telegraph, telephone; radio; electric dynamo; films; TV; computers and appliances. A century after the adoption of electricity as a technological resource and its incorporation into human activities, even a momentary lack of electricity can cause serious social and economic disruption. The electric power gave flexibility and other dimensions to all types of motors and machines.

The introduction of cutting-edge technologies and the technological convergence observed nowadays can, somehow, be considered similar to the introduction of electricity into people's daily lives a century ago. The introduction of these new technologies undoubtedly brings great changes in labor relations at the service, industry and agriculture sectors, including aspects related to safety, health and working environment, as well as the technological unemployment generation (NSF, 2001).

The adoption of new technologies, from a historical perspective, has resulted in higher labor productivity, higher profits for entrepreneurs and conversely, increased unemployment rates for workers (Brown & Campbell, 2005).

In the early days, whenever new technologies resulted in new occupation forms, part of the contingent of skilled manpower previously waived was absorbed. This occurred in agriculture, industry and services sectors. Today's situation is not different. New occupations and consequently new positions are being created mainly in IT and knowledge generation sectors. Accordingly, the current technological changes demands for very skilled workers. New job positions, nevertheless, are out of reach of the capabilities of most workers and, as a consequence, an intensification of technological unemployment is being observed (Brown & Campbell, 2005).

Nonetheless, one difference has emerged with the latest technological advance. Not only machines are replacing the 'muscle power' in manual work; computers are now replacing lightweight and 'intellectual' work. New technologies, including the so-called cutting-edge technologies, are fast replacing human beings in virtually every sector–from agriculture, to retail and government (Rifkin, 1995; Walker, 2014).

It is believed, by society as a whole that new technologies' emergence increases labor productivity and reduces costs, which, in turn, increases consumption, as products become gradually cheaper. As a consequence of consumption increase, more job positions are created. This is the model currently adopted by several capitalist countries.

The effects of technological development on unemployment generation can be clearly observed by analyzing two different examples from very distinguished moments in time. Whilst in 1904 it took a man 1300 hours to build a car, in 1932, 19 hours only were required (Rifkin, 1995). More recently, due to the automation in agriculture, a sugar cane harvester can replace approximately 80 workers during harvesting season.

It is remarkable to mention that, despite the invocation of 'new technologies' is an obvious and appealing call, general public and the media continues to ignore that some of these technologies have been already introduced into the market as very innovative products or as being part of new products.

Earlier and even nowadays, scientific and technological development are, in its first moment, very well accepted in societies (by the general public), once it is always associated to improvement of life conditions. However, it is important to point out that technological development has sometimes brought social and environmental adverse outcomes, as the detonation of the atomic bomb or the nuclear power plant accidents, in Chernobyl and Three Mile Island.

Analysts at that time admitted a technological utopia possible, whereas denied the possibility of the social one. This is a hegemonic vision also nowadays. The contemporary technological utopia is based on biotechnology and nanotechnology which pose serious unknown risks to workers and society as a whole (Kurzweil,

2005). Some researchers believe that the nanotechnology is the flagship of the Fourth Industrial Revolution. Table **1** presents, briefly, the relation among Industrial Revolutions, scientific development and impacts on workers and society.

THE SPEED OF SCIENTIFIC AND TECHNOLOGICAL DEVELOPMENT IN TIMES OF TECHNOLOGICAL CONVERGENCE

There are several contemporary authors that seek to address today's exponential phase of the scientific and technological development in capitalist societies. Among them, we chose to present a summary of Ray Kurzweil's point of view about this specific topic. The exponential increasing speed of the technological development was analyzed in his book *The Singularity is near: when humans transcend biology* (Kurzweil, 2005), using 110 of 652 pages for this purpose. Kurzweil articulates the nature of human life and how it could exist in a situation where men and machine knowledge have already merged. The author postulates a time when a growing and intimate collaboration between our biological heritage and a future that transcends biology. Kurzweil (2005, p. 4) predicated that human beings have the ability to "understand their own intelligence, access their source code and then -if they will- revise and expand it". The technology, according to his point of view, is the '*modern magic*'. Technological evolution can be considered as the continuation of biological evolution. He refers to this as Singularity, understood by the author as

> "a future period during which the pace of technological change will be so rapid, its impact so deep, that human life will be irreversibly transformed… this epoch will transform the concepts that we rely on to give meaning to our lives, from our business model, to the cycle of human life including death itself" (Kurzweil, 2005, p.7).

The author postulates that nanotechnology will enable the manipulation of physical reality at the molecular level, thus allowing the production of almost any engineered material from inexpensive basic ones, and will ultimately, turn even death into a soluble problem. Nanolevel designed robots, for instance, will have substantial roles within the human body, including reversing human aging to an extent not imagined or possible to be reached before.

Table 1.   Industrial revolutions in Capitalism: some characteristics.

	First	Second	Third	Fourth
Start	1780	1913	1975	1989
Leading country	United Kingdom	United States	Japan	United States
Flagship	Textile industry (cotton)	Automobile industry	Automobile and electronic industry	Converging technologies; cutting-edge technologies
Paradigm	Manchester	Ford	Toyota	Smart products and processes
Devices	Spinning Jenny machine, steam-driven machines, railroads, cotton ginning machines	Combustion engine, oil, petrochemical	Computing, CNC machines, robots, integrated systems, telecommunications, new materials, biotechnology	Machine-to-machine communication, sensor based devices, smart devices, remote machine operation
Organizational basis of production	Fabric production, waged labor	Series production, assembly line, rigidity, specialization,	Flexible production, U-shaped layout, just in time, total quality,	De-centralized modular systems; flexible production; bottom-up production; molecular manufacturing.
Work	Semi-handmade, qualified, heavy-loaded, unhealthy	Specialized, fragmented, intense, repetitiveness, unhealthy, hierarchical	Multi-task, integrated, team work, very intense, flexible, stressful, less hierarchical	Collaborative, creative, technological based, highly demanding, specialized
Investments	Low	High	Higher	Highest
Market structure	Free Competition	Monopolistic, strong verticalization	Monopolistic, strong horizontalization, outsourcing, large trading blocs development	Tendency to (intellectual) monopoly; strong patent-based development
Scale	Local, national, international	National, international	International, global	Global (impacts will cross borders); transnational corporations; local products International, governance and regulation are still needed
Productivity	Great increase	Great increase	Greater increase at breathtaking pace	Changes in the paradigm of productivity; vastly accelerated products improvement, strong customization under the conditions of high flexibility (mass-) production
Employment	Strong growth, especially in industries	Strong expansion mainly in large industries	Strong retraction, especially in industry; part-time, precarious, informal job	New job positions, new skill demands; increasing technological unemployment rates
Workers' reaction	Concern, rage against machines, cooperativism, first Unions	Concern, unions' strengthening, Social gains (wages, welfare, working hours, collective agreements)	Concern, de-unionization, fragmentation, tendency to the conflict or assumed partnership	Concern regarding new health and environmental hazards, new knowledge required; long-last learning needed.

Adapted and translated from CEFET-SP (2000).

According to him, the technological development is accelerating and expanding at an exponential pace. The acceleration of the paradigm-shift rate, in turn, is beginning to reach the '*knee of the exponential growth curve*', which is the stage where the exponential trend becomes noticeable to everyone (Kurzweil, 2005, p. 9). From this point, he also believes that future no longer resembles the past, and this is the change that makes difference in the new paradigm.

As a consequence of the exponential technological advances, there is also the emergence of new risks and associated uncertainties, which poses a unique challenge not only to consumers, but mainly to those workers involved in production of technological enhanced goods and to the environment as well. This is the remarkable change that will make difference in times of technological convergence.

WAYS TO LOOK AT RISK

Daily, scientific and technological research is being constantly developed worldwide. It necessarily implies the existence of risks and/or the possibility of risk existence. Regarding this, two aspects should be considered. The first one is the physical risk itself that impact or may impair human's (especially workers) and other species' health, as well as the environment. The second one is how these risks and possibilities are perceived by the general public (workers, consumers, society).

Flynn, Bellaby & Ricci (2006) analyze risk in terms of meanings. According to the aforementioned authors, risks are classified into three general types. Risk Type I is associated with its random statistical chance to occur (1:100,000 for example). In such cases, the probability of occurrence is based on previous evidence that this risk exists and is known. There is a risk basis which is known rationally and statistically treated. Risk Type II is more associated with taking decisions, when the potential consequences are based on, or depend on a future (not yet known). These risks are expected but can turn to be different than expected. Uncertainty in this type of risk is not merely statistically evaluated and risks, unlike Type I, are not known. Risk Type II travels a path into the unknown.

In such context, the higher the uncertainty, the higher the risk. Then, the main goal is to reduce uncertainty.

Regarding risk type II, in his paper *Managing and Communicating Scientific Uncertainty in Public Policy*, Brian Wynne (2011, p.2) states that:

> "The different approaches to uncertainty in public policy derive from correspondingly different implicit understandings of the nature of scientific knowledge and its relations with public policy. [...] The management of uncertainties in and around scientific knowledge for policy-making cannot be properly handled unless the larger questions over this wider terrain are borne in mind".

The author reinforces that what is meant by uncertainty is itself a matter of deep, though often implicit confusion. However, he presents distinctions clearly recognizable in the available literature, for the following concepts (Wynne 2011, p.5):

"**risk:** known damage and probabilities;

uncertainty: known damage possibilities but no knowledge of probabilities;

ignorance: unknown unknowns (second-order uncertainty);

indeterminacy: issue and conditions, hence knowledge-framing open; maybe salient behavioral processes also non-determinate;

complexity: open behavioral systems, and multiplex, often non-linear processes so that extrapolation from robust data-points always problematic;

disagreement: divergence over framing, observation methods or interpretation. Questions of competence of parties;

ambiguity: precise meanings (hence salient elements) not agreed, or unclear".

Risk Type III, on the other hand, refers not to rational expectations neither to reduce uncertainty by contingency plans. It refers to how people perceive the constitution of a threat. Flynn *et al.* (2006) also states that perception of risk can be influenced by both ideology and culture; and that while ideology can be defined as common interests that differentiate groups or classes, culture reflects these groups' history.

In brief, it can be said that in Type I, exposure to hazards is never probabilistically zero and, risk assessment typically involves framing. Science must focus on the readily observable and measurable events. In Type II, the unexpected might always happen, and those responsible for risk management restrict their vision of the future according to the interests they are allied with. In Type III, assured threats (or assured safety) seem certain to those who believe in them. It depends on ideology, culture and personal experience of those who are suffering the impact of technological risks (Flynn *et al.*, 2006).

An interesting contribution to the risk perception process (risk type III), with regard to nanomaterials, is presented in the literature review entitled *Risk perception and risk communication with regard to nanomaterials in the workplace* (European Agency for Safety and Health at Work [EU-OSHA], 2012). After a review of the theoretical framework about risk perception, the report indicates that (p. 9):

> "[…] the factors that trend to determine the level of concern are dependent on both the positive and negative effects a technology may have and the values or emotions of an individual. Their evaluation of risks takes place in a complex decision-making process where objective and scientific knowledge is not the key factor in risk assessment or acceptance leading to differences in opinion about the acceptability of the risk.[…]"

It is well known that many factors can influence risk perception and acceptance. Most of them depend on the qualities attributed by the perceiver to the source of risk. Psychometric studies, carried out by Slovic (2000), explain why some characteristics are more concerning than others to a given population; some of them described in Table **2**.

Table 2.   Factors affecting risk perception.

Factors	Low Risk Perception Factors	High Risk Perception Factors
Benefits	High benefits	Low benefits
Choice of exposure	Voluntary	Involuntary
Type of risk	Chronic	Catastrophic
Familiarity	Old risk	New (unfamiliar or novel source)
Catastrophic potential	Common – a risk that people have learnt to live with	Dread – a risk that evokes an emotional fear response
Visibility of exposure	Visibility	Invisibility
Individual control	Possible	Not possible
Origin	Natural source	Man-made
Risk management ability	No possibilities a priori	Lack of effective measures
Knowledge about risks	Known to the individuals expose (possible precaution)	Not known to the individuals exposed
Uncertainty	Known to science	Not known to science
Manifestation	Immediate or reversible damage	Delayed or irreversible damage
Damage	Definitely not fatal	Definitely fatal
Fair distribution of damage	Equitably distributed	Not equitably distributed
Damage visibility	Anonymous victims	Victims identifiable
Victims	Adult males	Children and women
Social or scientific status	Consensus possible	Controversial

Source: adapted from Slovic (2000).

Essentially, the author observed that three main characteristics drive the risk perception: the benefits associated with the risk; the catastrophic potential; and the level of knowledge about risks (Slovic, 2000). He also observed that these factors are broadly similar throughout the world. It is important to keep in mind that such characteristics are also present in relation to nanomaterials. As conclusions of the literature review produced by the EU-OSHA (2012, p.65), it is interesting to highlight remarks regarding risk perception and acceptance of nanomaterials:

"[…] Lay people and some workers generally have minimal knowledge and understanding of nanomaterials and are therefore unable to reach an informed stance... the fact that materials, with essentially the same name, can have vastly different properties at the nanoscale can cause confusion and misunderstandings. People generally expect risk from dangerous substance to increase with quantity, whereas nanomaterials are often handled in smaller quantities, which can add to this potential confusion [...] As an emerging scientific area nanomaterials have several inherent characteristics (eg uncertainty, lack of familiarity, potentially delayed and irreversible health effects man-made) that are likely to engender concern, mistrust or fear. There is therefore potential controversy which varies with sector and application within that sector, for example medical applications for the purpose of curing disease generally meet with public approval whereas medical application for improvement human performance do not [...]. The huge scope, novelty, excitement about promised benefits couple with the uncertainty and low current level of understanding in a rapidly changing scientific field also poses significant for risk communication."

According to the above mentioned review (EU-OSHA, 2012), nanotechnology is, at present time, mostly framed in terms of benefits, both by the media and general public; risks are then underestimated. Despite that, based on the scientific material available, it is neither possible to state how safe it is, nor take decisions on how to proceed, once nanomaterial hazard characterization is still controversial and scarce. The context of uncertainty in aspects such as hazard characterization, toxicity and exposure levels, directly affect the assessment of risks arising from nanoparticles to both human health and environment. In such cases of uncertainties, the right risk/benefit balance of a given process, the adoption of the precautionary principle is recommended, in order to ensure a healthy and safe development and application of nanomaterials for workers, environment and society.

Such considerations are important warnings agreed by the authors of this paper: risk perception, evaluation and assessment are not uniform over various technologies, countries and time.

The great amount of uncertainty regarding nanotechnology is due to the struggle between capital interests and labor on the knowledge production. Therefore very little has been effectively concluded, which makes it difficult to decide where to apply public resources to the development of new knowledge. Such knowledge development could reduce uncertainties about the use and application of nanotechnology and its impacts on living beings and ecosystems.

Although stakeholders, such as workers and their representative institutions, consumers and the general public, may have manifested about it over the last decade, it is clear that many governments do not carry out a satisfactory public program regarding nanotechnology. In many countries, there are no initiatives in such area. Generally, the opinion makers tend to pass on the virtues of nanotechnology only, usually ignoring the possible impacts and uncertainties of nanoparticles and nanomaterials in occupational health and safety.

In this regard, Nick Bostrom (2002) presents a complementary (and diverse) view about risks in his paper *Existential Risks: Analyzing Human Extinction Scenarios and Related Hazard.*

In the history of humanity, risks have always existed, since people are born until the day of their death. It is quite likely that in the early days of the human species, the highest risks occurred during childhood, in which the individuals were extremely unprotected, due to – for example - physical limitations (as the complete lack of awareness, mobility and self-preservation instinct), mother-dependence to get food and inability to move as easily and quickly as animals from many other species. As the time went by, nevertheless, humans acquired some skills that allow survival by, for example, learning how to get food, defend themselves from natural threats (or hazards) and enemies, besides dealing with social risks of wars and/or disputes with others of the same species and social groups.

According to him, just more recently the humanity has started to understand the extent of the risk that exists as the extinction of the planet due to their own actions or by catastrophic events that may take place at any time. For hundreds of years, the risks that threatened mankind, individually or collectively, were derived only from nature itself, dangerous animals and diseases (Bostrom, 2002).

The emergence of the first technologies, arising from science advances, has brought significant improvement into the living conditions, as well as played an important role in the drastic reduction of the risks (or threats to the species) from the point of view of the individual in its social group.

But Bostrom (2002) goes farther and proceeds to analyze possible human extinction scenarios and related hazards, derived from the rapid acceleration of technological progress in recent decades. In his opinion, mankind may be fast approaching a critical stage in its evolution. In addition to threats of nuclear world war, the unregulated proliferation of transforming technologies, such as nanotechnology and artificial intelligence, represent extraordinary and unprecedented opportunities and risks. Perhaps, the future of humanity may be determined by the ways in which issues concerning technological development and its applications are being conducted and addressed.

The author establishes three dimensions to describe the magnitude of the risks: *scope*, *intensity* and *probability*. *Scope* is related to the size of the group of people that are at risk. *Intensity* is related to how much each individual in the group could be affected. *Probability* understood as the best current subjective estimate of the probability of an adverse outcome. Based on their scope and intensity, he also distinguishes six types of risks (Bostrom, 2002, p. 1).

Bostrom discusses more profoundly the aspects related to the sixth category of risk, called by him as "existential risk" (Bostrom 2002, p. 1). An existential risk is one where humankind as whole is imperiled, and it is defined by the author as "one where an adverse outcome would either annihilate Earth-originating intelligent life or permanently and drastically curtail its potential" (Bostrom, 2002, p. 3).

The identification of risks in this sixth category is, according to his point of view, a recent phenomenon. For this reason, existential risks need to be differentiated from others. Humanity is not ready for managing such risks. Intuitions and existing strategies for coping with risks are still strongly linked to the mankind's past experience, such as pointed by the author:

"dangerous animals, hostile individuals or tribes, poisonous foods, automobile accidents, air, road, rail, sea, landslides, floods, fires, Chernobyl, Bhopal, volcano eruptions, earthquakes, drafts, bank failures, assaults, World War I, World War II, epidemics of influenza, smallpox, black plague and AIDS. In fact, these types of disasters have occurred many times and cultural attitudes towards risk have been shaped by trial-and-error in managing such hazards" (Bostrom 2002, p.2).

Nevertheless, existential risks are a different kind of risk; they have never been observed before. There is nothing to prove their existence, which in turn, become hard to take as seriously as it would be (simply because they have never been witnessed before). Decisions are then based on the subjective sense (the core of the precautionary principle), and risk factors are usually characterized according to one's best current subjective estimate (Bostrom, 2002).

For a risk to be recognized as such, there must be reasonable evidence that there is a probability (even a subjective probability) of a harmful result (an adverse effect), even if later it is possible to establish that objectively there was no chance of an adverse outcome. But, just knowing or not knowing if something is objectively risky already indicates that there are risks in a subjective sense, converging to the philosophical concepts of Noumenon and Phenomenon.

Kant states that an observer can know *a priori* only the phenomenon (thing as it appears), but not the thing-in-itself (*das Ding an sich*) - in the Kantian language, the noumenon. One thing is the reality as it is, and another is the way it appears before an observer, as a subject of knowledge. Man cannot penetrate the reality in its essence (noumena); it is unknowable (never knowing things in themselves). Man can only understand how reality appears to be (phenomenon); all that is perceived are nothing but representations or appearances (Hanna, 2005).

According to Bostrom (2002), unlike other categories, existential risks cannot be one of trial-and-error, just because there was no previous opportunity to learn from errors. Instead a reactive one, a proactive approach must be adopted. It is necessary to foresee new risk factors and to take preventive actions, bearing the costs of them.

The first existential risk faced and caused by mankind was the detonation of an atomic bomb, and sequentially, the build-up of nuclear arsenals in the USA and the USSR. In both cases, at that time risks could be qualified as *global* and *terminal*. By the time it was posed as risk, there was no past experience, no objective probability for harmful consequences, nor strategies for coping with risks that might have an unprecedented magnitude, which could destroy human civilization (Bostrom, 2002, p. 3).

The deliberate or accidental misuse of nanotechnology can be also classified as an existential risk to humanity. The extinction of intelligent life on Earth, for example, might be caused by the intentional release of nanobots/nanoweapons into the environment (Bostrom, 2002).

The prominence given to risk stocks is focused on the fact that, until recently, the development of technologies was perfectly framed and limited in their potential damage as *local* or *global* effects, considering that risks involved were *bearable* but not *terminal*. The longer-term danger of uncontrolled nanotechnology proliferation is an existential risk and can be understood as a threat to the entire world. Recognizing such existential risks suggests that is advisable, among others, the creation of a coordinated global strategy for the implementation of integrated security systems. National sovereignty must be respected, but taking into account the welfare of future generations, and not as an excuse for failing to take preventive measures against potential existential risks, such as nanotechnology (Bostrom, 2002).

WHY NANOTECHNOLOGY

Nanotechnology encompasses an extraordinary diversity of technological approaches currently under development. Even so, it has created a relatively new industry. The goods being produced and introduced into the market are usually innovative and unfamiliar to most consumers (Weekly, 2008). It was chosen as reference to this chapter, due not only to the authors' experience, but also by the fact that nanotechnologies were identified as an important source of known and unknown risks, as an existential risk, and due to the increasing uncertainties posed by these technologies regarding unique challenges in occupational health and safety, as well as environmental impacts.

According to the report *Green jobs and occupational safety and health: foresight on new and emerging risks associated with new technologies by 2020*, published by the EU-OSHA (2013a), nanotechnologies and nanomaterials are the most cited technologies with potential impacts on OHS by 2020. They occupy the first position among 26 pre-identified technologies, followed by wind energy and biotechnologies. The report refers to the generation of so-called *Green Jobs* in the European Union, taking into account that even this kind of job generates risks and uncertainties especially in the field of occupational safety and health.

Nanoscience and nanotechnology can be defined in several ways. Even though there is some ambiguity about what is nanotechnology, three main concepts define it: (i) its nanoparticles have size between 1 and 100 nanometers; (ii) it was designed, manufactured, or created at the nanoscale; and (iii) it shows specific properties that are scale-dependent. The National Science Foundation (2000) defines nanotechnology as:

"Research and technology development at the atomic, molecular or macromolecular levels, in the length scale of Approximately 1-100 nanometer range, to provide a fundamental understanding of phenomena and materials at the nanoscale and to create and use structures, devices and systems that have that novel properties and functions because of their small and/or intermediate size. The novel and differentiating properties and functions are developed at a critical length scale of matter typically under 100 nm. Nanotechnology research and development includes manipulation under control of the nanoscale structures and their integration into larger material components, systems and architectures. Within these larger scale assemblies, the control and construction of their structures and components remains at the nanometer scale. In some particular cases, the critical length scale for novel properties and phenomena may be under 1 nm (eg, manipulation of atoms at ~ 0.1 nm) or be larger than 100 nm (eg, nanoparticle reinforced polymers have the unique feature at ~200-300 nm as a function of the local bridges or bonds between the nanoparticles and the polymer)".

For the purpose of this chapter, the following definitions were adopted:

- Nanoscience corresponds to the research phase that aims at understanding the effects and their influence on the properties of materials at the nanoscale.

- Nanotechnology exploits such effects to build structures and to assemble devices with novel properties and functionalities and thus liable to produce marketing goods.

NANOTECHNOLOGY ISSUES CONCERNING HEALTH AND SAFETY AT WORK

As described in Health and Safety Executive homepage (HSE, 2014), "nanotechnology is an emerging field and it is expected to be the basis of much technological innovation in the 21st century. However, along with such innovation there come uncertainties as the unique properties of engineered nanomaterials pose an occupational health and safety risk". Ironically, the potential benefits that such materials can bring arise from these same unique properties Due to the novel characteristics, engineered nanomaterials have found innumerable applications in cosmetics, clothes, electronics, biomedicine, aerospace and computer industry (De Jong, & Borm, 2008; Mansoori & Soelaiman, 2005).

New and emerging technologies imply in new and emerging risks, which ideally should be anticipated in order to identify and avoid impacts on health and environment. The annex 6 of the aforementioned report (EU-OSHA, 2013a, p.179-191) presents several technologies divided into four main categories, as listed below:

- Renewable energy technologies, such as wind and solar photovoltaic energy;

- Fossil fuel technologies, such as carbon capture and storage and clean coal energies;

- Other energy technologies, such as nuclear and hydrogen cells;

- Non-energy technologies, such as biotechnologies, green chemistry, novel materials, convergent technologies, waste management and nanotechnologies and nanomaterials.

Nanotechnologies and nanomaterials present a very wide range of potential applications, including, engine additives and new materials used in construction (e.g. nanocoatings and nanopaints transforming solar energy into electricity). Industrial applications of nanotechnology are expected to grow very rapidly. From 2011 to 2020, worldwide production is expected to reach 50,000 tonnes per year (EU-OSHA, 2013a).

Nonetheless, nanotechnology engineered materials introduced new health hazards for workers involved in development, processing, maintenance, and recycling of nanomaterials-containing goods (EU-OSHA, 2013b). There is also potential for unexpected environmental impacts and thus, detailed investigation of the material properties is needed at all stages of the product life cycle (Lowry & Casman, 2009).

Materials in nanoscale exhibit significant changes in their properties when compared to the same materials at larger scales. This is mainly because nanoparticles have a smaller size and a much larger surface area than the same amount of a given material in a bulk form.

As the size of a particle is decreased, the number of particles per unit mass increases exponentially, and the surface area per unit mass also increases, with more atoms on the surface of the particle, which increases surface reactivity. The smaller particle size and larger surface area results also in changes in other properties such as color, solubility, catalytic activity, resistance, electrical conductivity, mobility (in the human body and the environment), chemical reactivity and biological activity (Kreyling, Semmler-Behnke & Möller, 2006a; Kreyling, Semmler-Behnke & Möller, 2006b). In other words, every reader must keep in mind that, in this specific case, (particle) size matters!

Whilst this reactivity and other improved/modified properties are useful to engineered materials with novel properties, a further consequence is that their biological properties may be enhanced or changed relative to the material in a larger particulate form (Warheit, Reed, & Sayes, 2009). In other words, the properties that make nanomaterial so attractive can also make them potentially toxic to humans and to the environment.

A material considered safe to be handled in larger size can easily penetrate human body through intact skin or respiratory tract when reduced to nanoscale size. From the toxicological point of view, nanomaterials can present enhanced toxicity or new toxicological properties not seen in bulky materials. Many variables can alter toxicity, such as size, shape, crystal structure, water solubility, surface area, surface coating, agglomeration, porosity and charge (Barry, 2008).

It becomes clear that the characteristics of the substances in larger size do not provide sufficiently reliable and understandable information about their properties at the nanoscale. The manipulation and deliberately manufacture of nanomaterials, that are different from anything that occurs in nature, can lead to unintended and even unknown consequences to the environment and health and safety of workers. These potential impacts are the ones that really concern (Yah, Simate & Iyuke, 2012).

It is well known that the toxicological effects of ultrafine particles are much more severe as their sizes are reduced. Ambient fines and ultrafines are associated with increased cardiovascular and respiratory events, including death, in susceptible populations. For example, the 1952s London fog mortality was related to the surface area of the particulate material. Nevertheless, little is known about the mechanism by which nanoparticles are absorbed, transported into the body, accumulated in different tissues and organs and eliminated (dose, dimension and durability should be taken into account as key factors contributing to toxicity) (Kreyling *et al.*, 2006a; Kreyling *et al.*, 2006b).

The knowledge area that studies nanomaterials and its impacts on human health and environment is called nanotoxicology and is in its first steps. This means that rigorous safety information is yet limited. So, although there is much excitement on nanotechnology, toxicity of nanomaterials still remains relatively unknown. Nanotoxicology is not progressing as fast as nanomaterials development. Due to the lack of toxicological data, there is a large gap in the knowledge regarding how nanomaterials exposure may impact and also impair life, health and environment as well (Barry, 2008; Chan-Remillard, Kapustka, & Goudey, 2009).

This gap of knowledge may hamper the ability to ensure workers' and consumer's safety as well as effectively regulate the production of nanomaterials. The lack of information is directly related to the uncertainties regarding the safety and health of workers. These uncertainties must be better explored and more robust strategies and policies must be developed, offering decent safe and healthy working conditions (Chan-Remillard *et al.*, 2009).

Despite the importance of nanotechnologies safety & health aspects, public funds addressed for nanotechnology knowledge production has mostly been focused on the science of production but very little on the science of impacts. Occupational safety and health seems to be of relatively low importance to governments and employers. They regard OHS issues only in terms of profit impacts only. It is imperative that such issues are undertaken in time, so that the pace of the development does not leave occupational safety and health matters behind.

According to Linkov & Satterstrom (2008), regarding the amount of resources applied over time and purpose, it is first applied on the production of knowledge, from the scientific and technological point of view. Safety & Health and Environmental data are generated secondly over time scale, later in time and in small volume. Assessment and regulation on these data are the last generated material, much later in time and in smaller volume as a consequence of lower financial resources supporting such activity.

Considering reduced public and private resources being allocated on the production of toxicological data in the field of nanotechnology, many questions remain unanswered, such as: what are the potential impacts on the workers' safety and health? What are the appropriate protocols for testing new products' nanotoxicity? What are the safety protocols for handling nanomaterials? Have proposed control measures already been effective? Have the effectiveness of respiratory protective masks, regarding nanosized fibers filtration been evaluated?

Other aspects should also be taken into account, and will be approached later on. Considering that traditional job positions could be extinct and others, created, an accurate assessment of the impacts of nanotechnology on job and social relations as well as its societal impacts should be carried out.

FOUNDATIONS FOR EFFECTIVE NANOTECHNOLOGY RISK ASSESSMENT AND MANAGEMENT

The understanding of the interrelationships among environment, human health and nanotechnologies undergo various conceptions. The authors believe that nanotechnology, in its development process, must be subject to some guiding principles that can oversee nanotechnology and nanoparticles.

The International Center for Technology Assessment (ICTA) along with the NGO Friends of the Earth, both from USA, and together with about 70 entities from different continents developed, published and signed the declaration entitled *Principles for the oversight of nanotechnologies and nanomaterials* (Nanoaction Project, 2007).

This document declares eight essential principles aiming at providing the ground for adequate and effective assessment and management of nanotechnology development, including those nanomaterials that are already in commercial use. Three of eight are directly related to the issues under review, reflecting interactions among environment, human health and nanotechnology and will be approached here.

The First Principle is called *A Precautionary Foundation*. It approaches the precautionary principle according to the statement: "When an activity threatens human health and the environment, precautionary measures should be taken, even when cause and effect relations are not fully established in a scientific way" (Nanoaction Project, 2007, p.4). In a synthetic view "no health and safety data, no market". Due to the lack of toxicological data of materials at nanoscale, which can differ considerably from their toxicity profile in bulk form, rigorous accurate and comprehensive pre-market safety assessments are required, taking into consideration such properties.

The Third Principle is called *Health and Safety of Public and Workers*. It is the second explained here and aims at protecting the health of the public and workers. Nanotechnology industries will employ approximately two million workers globally by 2015, according to the U.S. National Science Foundation (2001).

Due to its size, nanoparticle can cross biological membranes, cells, tissues and organs more easily than larger particle sizes. Once inhaled, these particles can pass through lung into the bloodstream. Nanoparticles can also penetrate skin and access the bloodstream. When ingested, nanoparticles may pass through the gastrointestinal wall and thus reach the circulatory system. Subsequently may adhere or infiltrate into various organs and tissues including brain, liver, heart, kidney, spleen, bone marrow and nervous system (Barry, 2008; Chan-Remillard *et al.*, 2009).

Workers at research, manufacturing, packaging, handling, carrying, use and disposing of nanomaterials are those mostly exposed to and most likely to suffer any potential harm. Despite the increasing number of workers in nanotechnology-related area, "no existing occupational safety and health standard specifically addresses nanotechnologies and nanomaterials, and there are no accepted standard methods for measuring human exposure to nanomaterials in the workplace" (Australian Manufacturing Workers' Union [AMWU], 2008, p. 8, para. 37). It is worth noting that the standard procedures known for the protection of workers must be revised to be suitable to nanomaterials use, when this is possible.

The Fourth Principle is called *Environmental Sustainability*. It is the third presented in this subchapter. It implies understanding that we should perform an assessment of the nanomaterial life cycle (manufacturing, transport, product use, recycling and disposal) before this nanomaterial is placed on the market. Once free in nature, products containing nanomaterials represent an entirely new class of contaminants. Therefore, new environmental impacts and damages may be expected (Lowry & Casman, 2009).

Reflections presented below are based on a paper published by Hannah & Thompson (2008), entitled *Nanotechnology, risk and the environment: a review*. It is a complementary view that corroborates to the adoption of the foundations described above.

Nanotechnologies already are and will be framed as environmentally friendly, being identified as promising in technology sectors such as fuel, energy, filters, environmental monitoring among others. Just to mention, nanoproducts are expected to enhance the performance of fuel additives, decreasing the CO_2 emission. Nevertheless, occupational and environmental exposures to engineered nanoparticles have not been fully characterized yet, nor understood. Only a limited number of engineered nanoparticles have already been analyzed to date, regarding occupational and environmental impacts (Hannah & Thompson, 2008).

The occupational and non-occupational routes of exposure to nanoparticles are already known and include inhalation, skin absorption and ingestion. Exposure to nanomaterial can occur *via* intentional or unintentional release, mitigation, and spillage or disposal of consumer products (considering also the likelihood nanoparticles uptaking by soil, plants, animals and water) (Hannah & Thompson, 2008).

Health hazards coming from exposure to engineered nanomaterials is not an easy task. Even if the structure, size and properties of nanoparticles are known, their interaction with the environment varies intensely, and thus, the traditionally used dose-response relationship does not apply. Current studies show that surface area is more important to dose-response than mass concentration (Yah *et al.*, 2012). As already said, nanoparticles behave differently from larger particles of the same material, into the human body. According to Borm & Kreyling (2004), toxicity of nanoparticles absorbed by inhalation is influenced by dose, deposition, dimension, durability and defense mechanisms. It will depend on the efficiency of these mechanisms, and all of them link toxicity to surface concentration. Another important aspect to be pointed out is that nanoparticles can be transported very rapidly through air, water or soil, which in turn, is directly related to the exposure measurement. The mobility in each media depends on size, charge, clustering, solubility, diffusion and deposition, among others, varying from particle to particle (Barry, 2008).

Generally speaking, there are two basic approaches regarding how the unique challenge posed by nanoparticles with regard to environmental and health risks to humans. Most scientists from all over the world, including Brazilians, heed the proposal that risks of nanotechnology must be studied case-by-case. After characterization of nanoparticle safety, a certain nanoparticle is characterized as insecure, control measures should be determined and applied. Thus, the risk management of nanotechnology involves measures to replace nanoparticles with less toxic ones, reduce the production and the exposure to these nanoparticles, as well as control the release of nanoparticles.

This risk-based approach is adopted by the USA government agency responsible for actions in nanotechnology. Founded in 2000, this agency is called *National Nanotechnology Initiative* (NNI). Nonetheless, studying risks of nanotechnology, case-by-case, is not feasible due to institutional, budgetary and human resources constraints, even in the USA and Europe. What will happen when thousands of different nanoparticles are in the market?

The second approach, already presented here as the First Principle for oversight of nanotechnologies and nanomaterials (Nanoaction Project, 2007), proposes the adoption of the Precautionary Principle, due to the uncertainty associated to the risks of nanotechnology. This principle, in its strong positioning (formulation in the original document), requires that no new technology be implemented, until a proof of safety has been provided. In a milder form, it requires precautionary actions that should be taken in the absence of scientific evidence of harm arising from nanotechnology. In such cases, the approval of products and processes should be delayed until more evidence of safety is provided.

Currently, USA, Brazil and some countries in Europe are adopting toxicity parameters of bulky materials of same composition to perform the risk assessment of nanoparticles, as a minimum regulation standard. From the toxicological point of view it is not suitable, once nanomaterials present enhanced and new properties compared to larger particles. Health hazards (including toxicity parameters) arising from nanoparticles must be clearly defined and characterized, before risk assessment performance. It must be taken into account that, in such case, risk assessment is a process to determine the risk of injury or illness associated with an identified hazard. It is based on information regarding physico-chemical properties of the substance, its nature, its health effects and how it is being manipulated and/or processed in the workplace. If the hazard is not identified or, prior to that, the sense of what is hazard is not defined, the risk assessment will not succeed (Ribeiro & Pedreira Filho, 2006).

An occupationally and environmentally sustainable management of nanomaterials must address such failures, in order to better regulate the production and commercialization of engineered nanomaterials. Due to the potential of nanotechnologies to transform both social and economic scenario globally, public participation in the deliberative and decision-making processes becomes essential, similarly to what happened with genetically modified products. Certainly, there will be a time when engineered nanoparticles will be present and will possibly prevail not only in several production processes but also in the environment (Hannah & Thompson, 2008; Nanoaction Project, 2007).

SOCIETAL IMPLICATIONS FROM NANOTECHNOLOGY

In addition to occupational and environmental risks, nanomaterials present broader societal and economic concerns. Due to its characteristics of producing new products in new ways, nanotechnology may be considered a disruptive technology. Such new capabilities bring large-scale changes, which may result in socio-economic improvements or may create new problems (NSF, 2001).

At a first glance, new technologies may cost more and be less effective, but as time passes by, they become cheaper and better, displacing earlier technologies. Specific industrial sectors (e.g. petroleum, agriculture) might be deeply changed, or even economically reduced to insignificance. Manufacturing windshield wipers for automobiles in Germany is a real example of such disruption. As car windows glasses are now covered with nanofilms to prevent adherence of particulate material or to improve water repellence, wipers will go obsolete. Alongside, industries producing such nanomaterials will dramatically grow.

The manufacturing industries profile changes result in organizational changes, with new division of labor and new competence demands to new job positions. In such situation, some workers might become redundant, while others may prosper. Those with the resources and adaptability to be requalified may succeed, while others may not make the transition successfully. This occurred during the deployment of other technologies, with computer for example, and is expected to also occur with nanotechnology (NSF, 2001).

Professional qualification will be dramatically impacted due to new skills and knowledge required by the market. Because much previous information will become useless, workers will have to be re-qualified to still integrate the workforce. This will pose major challenges mean to employers, employees, trade unions and will require deep changes on educational system. For instance, workers will have to be trained to use sophisticated soy harvester machines, once they bear nanotechnology-based sensors.

As another possible societal impact arising from this technological evolution, the retirement system will be directly affected, once a great number of job positions will be closed and workers headcount will drop dramatically. This will result in cutting down incomes, ruining *social welfare* and affecting pensioners negatively (Bainbridge, 2007).

Substitution of raw materials by new nano-engineered products may also impact socially and economically. Nanotechnology might develop new raw materials that may disrupt markets for pre-existing commodities, as such new nanomaterials gain widespread use. It implies potentially devastating consequences for the economies of natural commodity-supplying countries under development, whose most Gross Domestic Product (GDP) depends on mineral and agricultural exports (e.g. copper, iron and soybeans). If the above takes place, these countries could face dramatic financial problems once their trade balance would be affected. It would worsen the economic gap between poor and rich countries (ETC Group, 2005).

Poor countries will become poorer and consequently more likely to be increasingly dependent on richer and technologically developed countries. They will not probably groove the benefits from environmentally and friendly nanotechnological solutions, products and processes. Instead, cheap and outdated technology will be exported to them, as suitable and affordable solutions according to their current status on global economy (ETC Group, 2005; Invernizzi & Foladori, 2005).

At last, nanotechnology can be seen as a multipurpose technology that can broadly impact the life on Earth. Regarding global security issues, it must be kept in mind that military advances enabled by nanotechnology can be used to maintain worldwide peace. However, they can pose a substantial risk to mankind, if used for purposes of aggression and domination. An arm race based on nano-weapons could destabilize the current military balance forces; additionally, as become cheaper and widely available, nanoscale weapons could expand threats from terrorists or paramilitary groups (Bainbridge, 2007).

The societal implications of nanotechnology may be of a great scope and variety, presenting a highly diverse set of novel and technical possibilities. It can bring improved solutions to many environmental problems, as well as to human condition (health concerns included), assuring a status of global security and welfare.

For this reason, it is crucial to address both short- and long-term impacts, including ethical and social ones, intended and unintended consequences and effects of its use. It must occur at each stage of the development process (e.g. that could lead to redistribution of political influence as a consequence of redistribution of economic power).

A wide range of relevant actors with different interests are involved in this nanotechnological development. The interconnections among interested parties, which can produce, purchase or even fund nanotechnology, are responsible for shaping the societal impacts of nanotechnology. The allocation of public resources on studying the impacts of nanomaterials/nanotechnology on health and environment compared to the amount allocated for production of novel nanoengineered materials set an example of how government actors play a substantial role in developing a novel technological area.

It is a complex interplay of technical and social factors; the same technology that could be the basis for a new Technological Revolution (a new Industrial Revolution) must be developed in a way to avoid a situation where those who participate in the *nano-revolution* will take benefits from it, and those who do not, may find increasingly difficult to access the technological wonders that it engenders. Integrating actors from different areas can contribute to a healthy development environment where technological innovation and social aims are compatible and mutually self-reinforcing.

WINNERS AND LOSERS

Scientific and technological development has been the basis on which *Capital* has increased labor productivity - which implies on each worker producing more goods per time unit, and earning same or lower wages - and consequently increasing the *Capital* reproduction.

As this process was intensified with the globalization, there is an increasing production restructuring in various industries (based on science and technology development), resulting in the so-called 'technological unemployment'. It has been growing in several capitalist societies (Rifkin, 1995; NSF, 2001).

As a consequence, changes have been observed in the work organization, such as: downsizing; job losses; and changes in the employment relationship. The employment relationships are moving away from a long-term employee-employer relationship to multiple jobs, part-time jobs, or work at home. Substantial numbers of workers are now contracted workers, contingent, temporary, leased workers, day laborers and independent contractors. Qualifying employees has been deeply affected as well; life-long learning in this field is now imperative (Papadopoulos, Georgiadou, Papazoglou, & Michaliou, 2010).

The intensification of workload and working hours (for those who still have a job) is also a consequence. Besides, occupational health and safety conditions are being degraded. Such changes are directly related to the uncertainties associated with these new production processes.

As already previously mentioned, public and private funding resources are mostly allocated for the production of new nanoengineered materials (science of production), whereas the studies on occupational health and safety impacts (science of impacts) are let aside. It results in increasing uncertainties associated with the process, ending up affecting unevenly workers health and working environment (and society in general).

Thus, both workers in particular and society as a whole are the actual losers in this context of technological convergence adoption, (nanotechnology, biotechnology, information technology, and cognotechnology) and its unevenly distributed uncertainties.

The society is impacted in several ways. Firstly, the allocation of public resources is designed almost exclusively for the science of production. Secondly, there are no specific regulations on nanoproducts already being commercialized. As a consequence, uncertainties associated to the processes and products are passed on to the final consumers. Most of the negative impact, considered as capitalism's process externalities, is then addressed to the society, who pays the highest price.

Thirdly the dominant view, that technologically improved products become cheaper and widely distributed does not take place in oligopolized and monopolized sectors, such as nanotechnological-based industry. In such context, bearing specific technology destroys market competition. The exclusiveness of holding expertise and patents is one of the factors that contribute to avoid competition.

The *Capital* itself is the only one which benefits, assuring its production and reproduction, regardless of the negative consequences towards society, which are: the worsening occupational safety and health conditions and working environment and uneven distribution of risk uncertainties, the so-called process externalities.

We must point out that currently, in several countries, the enterprise productive restructuring, based on new technologies, is supported by Nation-State policy. New technologies, innovation and competitiveness have been the magic slogan that enables capital reproduction in times of globalization.

Nevertheless, the same nation-state has to create jobs and to assure the creation (or the maintenance) of human capital formation, investing money in workforce education, training and retraining. By doing so, it will cut down social inequality, allowing access to technological innovation to the whole population, that is, cutting down the process externalities.

PERSPECTIVES FOR WORKERS: THE ROLE OF UNIONS

Once the authors of this text are Brazilians, largely experienced on the field of world of work, Brazilian problems will be mostly approached in this subchapter.

In all cases of social and political discussions that have come along with the adoption of new technologies in capitalist economies, workers have tried to gather together in order to defend common interests, i.e. minimize losses and the uneven distribution of profits, ensuring that the benefits arising from technological development are reaped by everyone.

In view of the technological improvements witnessed in the First Industrial revolution, workers raged against machines. During the Second Industrial Revolution, workers organized themselves and unions were developed to give the workers a voice in the workplace to struggle for their rights, such as better working conditions, and virtually for a better and alternative society. As the Third Industrial Revolution, unions experienced some constraints, posing challenges for unions struggling worldwide. Despite that, workers mobilization for unions' continued relevance even under neoliberal global capitalism, keeping the struggle for improvements at workplace, their rights and values.

As for nanotechnology, it would not be different. Despite the growing evidence that nanomaterials poses unique health, safety and environmental risks, no country has introduced specific regulations addressing nanotechnology and nanomaterials (Del Castillo, 2010). Most of them rely on regulations that had not been designed to protect workers from nanosized materials. Unions present an important role, claiming for regulation, social and ethical control over the use of nanotechnologies in the working environment.

Throughout the world, workers organized in unions, intercontinental confederations or national associations have claimed positions regarding nanotechnology in recent years, also seeking national development (Australian Council of Trade Union Fact Sheet, 2009; European Trade Union Confederation [ETUC], 2008a; 2008b; 2010). The paragraphs below will show a summary of reports by the major national and international unions and confederations. It is not intended to intensely discuss the topic that follows, but to provide essential information for a critical thinking regarding the workers perspectives in a scenario of uncertainties and risk extended risen from new technologies, such as nanotechnology.

The International Union of Food, Agricultural, Hotel, Restaurant, Tobacco and Allied Workers' Associations (IUF) and the TUCA Trade Union Confederation of the Americas (TUCA-CSA) are the main protagonists of workers' struggle for the regulation, democratization and social control over nanotechnologies globally (Intercâmbio, Informações, Estudos e Pesquisas [IIEP], 2008).

IUF has been working on that since 2006, when their Latin American Regional Branch held its 16th Regional Conference. In 2007, during the IUF 25th Congress, nanotechnology was intensely debated and all the participants launched a call for strong, comprehensive regulatory oversight at all levels of nanotechnology and its products. IUF supports the adoption of the Precautionary Principle, claiming to international institutions, such as the *World Intellectual Property Organization* (WIPO), to suspend new nanoparticles patents until their potential risks posed by nanotechnologies can be effectively understood by all the social actors. IUF expects *World Health Organization* (WHO) and *International Labor Organization* (ILO) to study short- and long-term nanomaterials effects and impacts on occupational health, safety, environment and workplace thoroughly and also expects *Food and Agriculture Organization of the United Nations* (FAO) to update the *Codex Alimentarius*, which must encompass the use of nanotechnology in food and agriculture.

When unions of Americas began to intervene in nanotechnology debate, TUCA-CSA had not been created yet. At that time, workers, represented by the *Inter American Regional Workers' Organization* (ORIT), joined many civil society organizations, engaged in a coalition to demand the creation of a nanotechnology nanomaterial inspection principles.

Since then - and from 2008 as TUCA-CSA, the largest regional workers´ organization in the Americas - the American trade unions have been struggling for such principles implementation, among which are government specific regulations regarding nanomaterials; workplace monitoring; risk prevention programs, toxicological studies and other initiatives to ensure workers health and safety; evaluation of nanomaterials life cycle to avoid environmental contamination; transparency in risk assessment and supervision that also requires labelling and cataloging nanotechnology products; and ultimately, the democratic public participation in all aspects of the debate, including active participation in decisions about research and uses of nanotechnology.

In Brazil, the *Central Única dos Trabalhadores* (CUT), *Força Sindical* and *União Geral dos Trabalhadores* (UGT,) which are unions affiliated with TUCA-CSA, discussed and got into consensus, writing up a report on demands regarding nanotechnology regulation and assessment (Intercâmbio, Informações, Estudos e Pesquisas, 2008). Such decisions meet 2007 ORIT demands.

Like international unions, Brazilian ones are concerned with the spreading of nanotechnology, with no proper notion of its consequences on workers' and consumers' health and on the environment. The direct impacts on job positions are another priority. CUT, Força Sindical and UGT have warned stating we are before a new industrial revolution, which will mean restructuring economic sectors and consequently closing job positions, changing work organization, demanding new workers skills.

The first Brazilian trade union that discussed nanotechnology was the *Confederação Nacional dos Químicos* (CNQ, a chemical union affiliated to CUT), in its 5th Congress, held in June 2007 (Departamento Intersindical de Estatística e Estudos Socioeconômicos, 2008). On that occasion, CNQ/CUT reported its concerns about impacts on workers' health and environment due to increasing nanotechnology application in several production processes. The produced report has proposed regulation of exposure patterns, toxicological and ecotoxicological studies on effects and impacts, development of analytical instrumentation for sampling, measuring and analyzing nanoparticles, risk assessment tools, as well as the creation of institutions to monitor nanotechnology development. From 2010 on, the CNQ/CUT has held seminars, has brought unions together from all over Brazil, to disseminate the issues involving nanotechnology and nanoproducts, regarding risks to workers' health.

In 2008, the *Federação dos Trabalhadores do Ramo Químico no Estado de São Paulo* (FETQUIM, a regional chemical trade union) was the first to propose a clause demanding rights to accessing information. Since then, the issue has been intensely discussed with workers and their representatives. Such clause was also included in a Collective Bargaining Agreement in 2012. Below we comment on the main demands from unions regarding nanotechnology.

In April 2007, a document was created and named *Workers' Agenda aiming at Development*. It comprises guidelines, policies and actions towards changes to foster development in Brazil. It is divided into four topics: inequality wealth concentration; unemployment and labor; state ability to foster development; and society participation and democracy.

In such case, development is understood as a process of reaching environmentally sustainable common welfare, respecting social, political and cultural diversity. As for labor relations, development is understood as the promotion of decent work conditions, including collective bargaining.

Labor Unions consider that the consolidation of nanotechnology as a Fourth Industrial Revolution can simply block the foundation of a sustainable development. Therefore, Labor Unions have been working on reaching consensus to find basis for common actions regarding nanotechnology and its impacts on workers' health, which meets the above mentioned *Workers' Agenda*.

Firstly approaching Labor Relations, Unions state that Brazil needs to generate more and better job positions. The introduction of new technologies must be accompanied by the requalification of the human capital to meet the new job demands and skills, otherwise social and economic inequalities will worsen.

Workers mobilizations and collective bargaining must then include clauses about the rights to information regarding new technologies and their applications in the workplace, environmental protection and workers health; employers' duties in the prevention of nanotechnologies impacts on workers; and personnel requalification. The collective bargaining agreements must then include clauses that make employers responsible for any possible damage to the workers' health and environment due to the use of nanomaterials.

The debate over State role is crucial to labor movement. The Federal Government nanoscience and nanotechnology programs are based on quantitative data, by growth in number of scientific and technological products in nanotechnology. It is also based on patent growth; the growth of national companies which incorporate nanotechnology products and processes, as well as the nanotechnology material export growth. Unions want the government to establish parameters before allowing funds for new technology development.

The government has allocated funds to identify nanostructured products for export. However, there are no resources aiming at studying impacts on workers' health and domestic environment. Brazilian government needs to define its role more clearly regarding regulating the use and sales of nanostructured products.

At last, some considerations about social control and citizen rights. Nanotechnology implementation in productive processes has taken place in main capitalist based countries, where we find multinational headquarters. In countries under development, such as Brazil, nanotechnologies often might be incorporated into productive processes of these company branches escaping social and state control. Because companies, corporations and businesses in general have grown dramatically nowadays and therefore can mean strong political influence, unions demand the Brazilian government to guarantee more control over the nanotechnology based production.

It must be clear that, in general and particularly in Brazil, workers are mostly uninformed regarding the meaning of nanotechnology and convergence technologies, their inherent risks and impacts on health, and also the current uncertainties and data limitations whether the safety procedures implemented are adequate or the protection measures adopted are sufficient.

Countries in general supporting such technological convergence do not seem interested in letting society as a whole to discuss with them. Despite of union efforts, governments have done too little seeking to inform the general public.

In times of technological convergence, where nanotechnologically improved goods are being produced and commercialized, workers and consumers are being exposed to the unknown potential risks of nanotechnology and uninformed about them. There is no information process driven to society as a whole or to workers specifically in order to raise awareness about nanotechnology consequences. As a consequence, the uneven distribution of knowledge can create a nanodivider between those who can benefit from the innovations and those who can neither afford nor access them.

PRODUCTION OF NON-KNOWLEDGE IN NANOTECHNOLOGY: POLITICS, ECONOMY AND PHYLOSOPHY

Sciences generate and bring to society not only knowledge, but also ignorance and non-knowledge concerning the possible health and environmental impacts of their innovations and consequent technological application.

Non-knowledge is a result of the growth of knowledge itself. The term *non-knowledge* indicates the general absence of knowledge, disregard of its contextual implication. *Ignorance*, on the other hand, implies the availability of knowledge; *uncertainty* implies the recognition of a lack of knowledge, as well as its further qualitative specification (Boschen, Kastenhofer, Marschall, Rust, Soentgen, & Wehling, 2006).

Environmental politics and research should take into account not only the known and well-defined risks and uncertainties, but also completely unknown, unanticipated and unrecognized effects that emerges with new technologies. The delayed recognition of health and environmental risks has already been debated by scientists in cases such as asbestos, and the report *Late Lessons from Early Warnings* (European Environment Agency, 2001) has given the following advice:

> "Acknowledge and respond to ignorance, as well as uncertainty and
> risk, in technology appraisal and public policy making".

In this regard, the International Risk Governance Council (2010) published the
report *The Emergence of Risk: Contributing Factors* focusing on general origin of
emerging risks. The report defines *emerging risks of a systemic nature* as "risks
that reach more than one country, more than one economic sector, and may have
effects across natural, technological and social systems" (International Risk
Governance Council [IRGC] 2010, p.5). Such risks are not isolated events, even
though they originate in different parts of the world. Emerging, systemic risks
may affect human health and safety, security, environment, economies, and at last
the societies' framework. Many systemic risks occur in contexts in which their
future evolution is difficult to control; uncertainties and gaps in knowledge are
some of the specific challenges faced by those involved in management of
systemic risks.

According to the IRGC (2010) report, these risks emerge from a common
breeding fertile ground, which in turn, is cultivated by the following 12 generic
contributing factors: scientific unknowns; loss of safety margins; positive
feedback; varying susceptibilities to risk; conflicts about interests, values and
science; social dynamics, technological advances; temporal complications;
information asymmetries; perverse incentives, malicious motives and acts; and
communication. Such contributing factors should be understood not as discrete
units, but complex, interdependent factors.

Among them, the first contributing factor to be presented and discussed here is the
Factor #1: Scientific unknowns. This theme is central to the authors of this
chapter and approaches scientific production of non-knowledge. The knowledge
(or awareness) of non-knowledge varies from full awareness of non-knowledge
(known unknowns) to complete unawareness (unknown unknowns). The term
'scientific unknowns' incorporates both 'known unknowns' and 'unknown
unknowns'. According to this report, the 'known unknowns' are equivalent to
uncertainties; in such cases the current state of knowledge does not allow for the
quantification or description of the likelihood, magnitude or even nature of
potential adverse effects. The 'unknown unknowns' refer to a situation where

even less (or no) knowledge is available; in such cases quantification and description of risk is impossible. Scientific unknowns can affect the estimation of the likelihood or severity of an emerging risk, thus having greater potential to amplify or attenuate emerging risks.

The scientific unknowns, in turn, can be classified as tractable (within the management control) or intractable (outside management control). Unknowns are considered tractable when a targeted research activity is expected to clarify the unknown or at least bound the possibilities within a useful range, with a high degree of confidence, and can aid in the identification and assessment of emerging risks. Intractable unknowns, on the other hand, are those that cannot be resolved in an appropriate timeframe to inform decisions about the emerging risks.

Scientific unknowns are framed in different modes in both science and society. Once its production is inherent to the knowledge production process itself, the ways of dealing with such non-knowledge will contribute to its better understanding, communication and management.

As well exemplified in the report (IRGC 2010), the mobilization of financial and human resources is a way to the elimination or reduction of uncertainties and knowledge gaps. Soon after scientists, both in France and USA, had discovered and isolated the retrovirus responsible for Acquired Immune Deficiency Syndrome (AIDS), rapid advances were made and the scientific unknowns, such as sequencing the HIV-1 genome or understanding modes of transmission, were solved. The pace of research over the period 1982-85 has been considered the fastest in medical history. Solving such scientific unknowns allowed for the development and broad use of blood tests for HIV, as well as anti-HIV drugs therapies. It is important to point out that there had been political decisions at different levels that provided financial and human resources, resulting in important breakthroughs between 1982-85, eliminating or reducing uncertainties concerning HIV/AIDS at that time.

Regarding nanotechnology, it is known that both industry and governments are investing massively in research and development of nanomaterials. On the other hand, produced non-knowledge is not being properly addressed. Assessment methodologies are not being developed in the same pace at which marketing of nanoproducts become reality. The produced scientific unknowns, such as health and environmental impacts, remain poorly studied.

Despite the attention given to the risk assessment, management and communication of hazards and adverse health effects which are known, particular attention should be given to what is not known regarding negative impacts of nanotechnology or any other technological innovation. Thus, it is imperative that societies deal with potential, and possibly not even fully known and foreseeable, risks posed by innovations and new products from nanoscience and nanotechnology. Additionally, risk assessment and management should also comprise the scientific unknowns, and how intractable those unknowns are likely to be.

The second contributing factor to be presented and discussed is the *Factor #5: Conflicts about interests, values and science* (IRGC 2010). It is related to the emergence or amplification of risks due to different understandings of the science as well as different values and interests of the stakeholders involved in risk management. It must be kept in mind that science, values and interests are entangled in mind and behaviors of people. If emerging risks involve conflicts, efforts should be made to pinpoint the underlying interests and values of involved parties. In such context, there must be a clear differentiation between the assessment of the science and the assessment of the values involved, and later, the evaluation of the risk acceptability, in order to guarantee an appropriate identification, prevention and mitigation of risk.

Governments, private sector, researchers and public have different knowledge and consequently opinions regarding nanotechnology development, as well as the incorporation of nanotechnologically enhanced materials into market.

Advances on nanotechnology field must be shared by scientific and industrial communities with regulatory agencies, civil society and the general public too. Lack of communication and understanding about nanotechnology application may impact societal perceptions negatively, which in turn, may affect the political and regulatory decision-making process. In such contexts, public confidence may erode, and risks can be amplified because its assessment and management may not be carried out taking into account the most accurate information and knowledge.

Another contributing factor is the *Factor #6: Social Dynamics* (IRGC, 2010). It refers to how social changes can lead to or attenuate potential harm. Identify, analyze and understand changing social dynamics is crucial on risk management. Societies are continually evolving. They may adapt to new technologies, economic forces or political ideologies (or may fail to adapt). Nowadays, globalization has significantly changed the risk scenario in many sectors. Despite the many opportunities it has brought, some believe that such benefits have not been equally shared; in their opinions poor countries have been left out and both social and economic inequality has risen in poorer countries. Development, poverty and inequality levels are important factors to contribute to a fertile ground for risks to emerge or amplify. In general, poorer or less developed societies are more vulnerable to emerging risks, as a consequence of lacking wealth, education, literacy or capital, due to their limited ability to access and understand information, which, in turn is crucial to the ability to identify, assess and manage risks.

Nanotechnology, in turn, can be considered as a disruptive technology that can change markets, production processes, geographical distribution of industries, as well as the workforce distribution. Societal impacts that can rise are closely related to the decisions regarding nanotechnology research policies, planning and directions, which may result in uneven or inequitable distribution of risks and benefits among different countries, according to their socio-economic profile.

Due to its disruptive characteristics, it is important to point out that the social scientists play a fundamental role on the better understanding of how nanotechnology can affect social dynamics and social behavior. Without accounting for that, many adverse effects will be neglected, and opportunities for risk prevention and mitigation will certainly be missed.

The last- and perhaps the most important in this chapter's context- contributing factor to be presented and discussed is the *Factor #7: Technological Advances*

(IRGC, 2010). In summary, it is related to the risk emergence when technological development is not accompanied by appropriate scientific investigations regarding health, economic, ecological and societal impacts. Technological advances, a recognized source of improved quality of life and welfare, can produce unwanted risks, especially when rapid diffusion of technology occurs without adequate risk assessment or monitoring of impacts. In such context, risks can be amplified when economics, policy or regulatory frameworks are insufficient. Countries around the world differ in the stringency of their requirements to screen for emerging risks through safety studies. As a result companies often seek approval of their new technologies, whose health impacts are yet poorly understood, in countries with weak regulatory systems.

Research, development and innovation in nanotechnology are far ahead of the policy and regulatory environment. Despite joint efforts made to improve risk assessment, management and governance, governments, industry and academia still face difficulties which need to be addressed. Such deficits include, for instance, lack of programs for monitor and control the impacts of nanomaterials in the workplace; lack of harmonization of risk assessment and management procedures; and fragmented regulatory structures, both nationally and internationally. Differences in national regulations also impact on international policy coordination, and could lead to a transferring risk to countries with weaker controls.

Nonetheless, it must be kept in mind that, discussions on knowledge/non-knowledge production and dealing; on how it is framed by the scientists and societies; and lastly, on regulation and governance of emerging risks arising from technological innovations, are intrinsically related with national social, economic and political conditions.

From the view of political-economic system, it is important to highlight that societies are predominantly capitalist all over the world. Basically, capitalism can be defined as the use of the *Capital* to obtain profit, which is mostly reinvested to obtain more *Capital*.

The first contradiction of capitalism, *Capital vs* Labor, refers to capitalism's tendency towards overproduction: in one hand, capitalism unlimited capacity to produce; in other hand, consumption constraining due to competitive pressures on capitalists to cut costs, mainly achieved by restraining wages and cutting their workforce (to extract surplus labor from working class) (Gonik, DeCarlo, Turner, & Thomas-Muller, 2010). Technological development, as mentioned before, speeds up the production, replacing workers by machines. It has been occurring in every sector of global economy. This, associated to low-paying jobs, contributes to the reproduction of *Capital* through unpaid labor. The same first contradiction also approaches the workers class struggle to fight labor exploitation. Such struggle also fights for risk reduction and better occupational health and safety conditions in working environment. The higher the capital power constrains workers organization, the higher the occupational risks will be.

Uncertainties play important role in forming risk scenario. They are closely related to how knowledge is produced and to who is funding such development. Referring to technological process innovation, uncertainties are being fought by investing more resources in knowledge production which results in new ways of manufacturing. This is called 'science of production', whose funds are mainly provided by governments. This applies to nanotechnology development worldwide, including Brazil, where uncertainties are permanently fought *via* governmental policies which are, in turn, considered a 'successfully adopted science and technology policy'.

Referring to occupational health and safety, uncertainties are scarcely fought. In EU and USA for example, the least possible is invested by governments to cut down risks (ca. 5%). In Brazil it is still worse: 1% only of available funds was invested from 2001 to 2011. The use of public investments on nanotechnology production are massively oriented towards production itself rather than preventing, understanding or fighting risks, which affects workers health and safety. This, therefore, encourages *Capital* reproduction using State resources.

The second contradiction of capitalism refers to the contradiction between *Capital* and Nature. James O'Connor was one of the authors that worked on this topic (O'Connor 1988; 1996; 1998; n.d.). According to him, capitalism suffers from a second contradiction due to Capital's addiction to growth. He means that (Gonik *et al.*, 2010, para. 2):

> "capital accumulation can be jeopardized by so fouling the natural conditions of production that it totally breaks down by raising the cost of production which arises from increasingly depleted raw materials and from the need to invent and develop substitutes; or yet by the state being forced to allocate increasing amounts of surplus value for restoring ruined soil, oceans and forests".

In other words, Capital limits itself by impairing its own social and environmental conditions, due to its incessant need to grow (Spence, 2000; Gonik *et al.*, 2010).

The second contradiction, thus, suggests that the cyclic reproduction of Capital takes much less time than nature reproduction cycle. Consequently to overcome the depletion of nature-based raw materials, industry has needed to produce synthesized ones. That enables the production of new technological knowledge, in order to produce substitute materials, feeding the reproduction of capital cycle in manufacturing processes.

Three key events, related to the technological development mentioned above, must be highlighted. Firstly, the manipulation of genes provided by biotechnology, that has allowed, for instance, the development of genetically modified crops, such as soybean, corn and canola, which have become more herbicide resistant. Secondly, the manipulation of materials at atomic level (organic and inorganic components) provided by nanotechnology that has allowed the production of biochips, miniaturized structures able to perform hundreds or thousands biochemical reactions simultaneously. Lastly, the making of engineered nanomaterials deriving from both nature and synthetic materials, also provided by nanotechnology, such as carbon nanotubes, which are tube-shaped materials made of carbon, and whose diameter is measured on the nanometer scale. Although being formed essentially by graphite sheets, such nanotubes can differ in length, thickness and number of layers. And, depending on these variations, carbon nanotubes can act either as metals or as semiconductors. Compared to other materials, they also show singular combination of stiffness, strength, and tenacity.

Synthesized in 1990's, due to the nanotechnology development, it was announced as material stronger than steel, unbreakably elastic, resistant to chemicals and lighter than aluminum. Genetically modified organisms, biochips and carbon nanotubes are not present in nature itself, but they are now part of ordinary life.

Alike James O'Connor, the environmental sociologist Allan Schnaiberg, also followed the Eco-Marxist School (Schnaiberg, 1997). According to his societal-environmental dialectic, ecological concerns will be overwhelmed by the desire for economic expansion. In other words, immediate economic growth will be maximized at the expense of environmental disruption, supported by political decisions. Then, to prevent health and economic disasters, governments will attempt to control the most extreme environmental problems only, which in turn, will give governments the appearance of acting more environmentally concerned than they really do. At last, upon reaching a severe environmental degradation, political forces would respond adopting sustainable policies, a process driven mainly by the economic damage caused by such degradation. At this point, economic engine would rely upon renewable resources and production and consumption would comply with sustainability regulations (Foster, 2005).

Schnaiberg's political capitalism, also known as the *Treadmill of production* (Schnaiberg, 1997), states that environmental disruption is caused by the acceleration of such treadmill. He also points out that the deceleration of this treadmill, that is, how environmental degradation might yield a breakdown in the acceleration-based treadmill alliance, can avoid such environmental disruption. Such deceleration could be achieved by state and working labor movements designing policies to fit the economy as reasonable as possible to its consumption requirements. Meanwhile, he argued a common alliance among state, labor and capital to support common economic growth.

We, as authors, agree with both O'Connor and Schnaiberg, pointing out that environmental problems caused by capitalism production modes, such as natural resources depletion, affects social relations, results in conflicts and do not contribute to a sustainable economy, and lastly, to a sustainable society. So, the following question remains: shall bio- and nano- technologies, already incorporated by Capital reproduction processes, overcome lack of natural raw materials and other environmental problems?

Nanotechnology supporters allege that any nanobased production processes use much less natural raw material, as well as energy, to produce the same thing. Consequently, it reduces the massive exploitation of high value natural resources, reducing social impacts and conflicts, and thus, overcoming the second contradiction of capital.

However, investments in knowledge production (science of knowledge) prevail and it is not accompanied by a proportional and fair investment in the science of impacts. It means that, regarding nanotechnology, the production of non-knowledge - the scientific unknowns, the uncertainties – grows dramatically implying potential adverse effects and impacts on environmental and workers health.

So, in order to avoid the depletion and massive exploitation of natural resources, both public and private economy sectors drive their efforts towards production of science, in detriment of being consumers-; workers-; or environmentally-friendly. This ensures the *Capital* reproduction on itself, once the restriction imposed by the lack of raw material is overcome by (nano-)technology. Such vicious cycle perpetuates the production of non-knowledge, since possible impacts on human health and environment are considered as a process externality, increasing uncertainties associated with nanotechnological development.

Costs from resulting damages and losses are transferred to society as a whole in several ways: workers are exposed to unknown hazards that may jeopardize health; population in general can be exposed to the same hazards if the surrounding environment is polluted by nanoparticles released by nanoproducts facilities; and at last, nanoparticles leach from products to accumulate in the environment. The lack of regulation and the lack of social control and participation in decision-making processes can also mean costs to the society.

The trends described above show that the private appropriation of public resources aiming at producing knowledge to sustain the capital reproduction leads to an uneven and unfair distribution of risks and uncertainties throughout the society (Rice, 2009).

Besides social; economic; and political issues, ethical implications posed by nanotechnology are equally significant. Nanotechnology is a rapidly growing field of scientific and technological innovation that can potentially impact and change the today understanding of how engineered materials interact with human beings and nature. It waves the flag of human welfare improvement by its own and also through convergence with other technologies, the so-called NBIC technologies: nano-bio-info-cognitive technologies (Roco & Bainbridge, 2003; Dupuy, 2009c).

Worldwide, the ethical concerns prompted the implementation of research programs on ethical, legal, and societal impacts of nanotechnologies; nanoethics was then born. Nanoethics, as a study of nanotechnology's social and ethical implications, is an emerging and already a controversial field (Ferrari & Nordmann, 2010; Bensaude-Vincent 2013; Nurock 2010).

Ethically, the first particular problem posed by nanotechnology is its extensiveness. So, it can be addressed, in terms of science, as a single and new technology, or by considering the extremely wide range of its potential applications. Some people believe nanotechnology to be merely a combination of several existing disciplines, such as chemistry, biology and physics. Others argue that nanotechnology is more than just the sum of individual parts and, for this reason, must be discussed as a particular field (Allhoff & Lin, 2006; Allhoff & Lin, 2007; IRGC, 2007).

Consequently, nanoethics has a broad range of meanings, according to different points of views (Franssen, Lokhorst & Van de Poel, 2009; Keiper, 2007). It can be understood as notion of responsibility, known as 'responsible research and innovation', which is mainly focused on the anticipation of the potential environmental, health and safety impacts, as well as the legal and societal impacts of the application of nanotechnologies. Or it can be restricted to prudence and be understood as merely rational risk management. Some reduce nanoethics to a moral cost/benefit analysis (Bensaude-Vincent, 2013; Dupuy, 2007).

It is not an easy task to specify what make a technology singular enough to need its own ethics. Does a change in scale justify by itself the emergence of a new branch of ethics, known as nanoethics? Is nanoethics really needed? How can the existence of nanoethics be justified philosophically?

Philip Ball (2003), science writer for *Nature*, says that

> "Questions about safety, equity, military involvement and openness are ones that pertain to many other areas of science and technology [and not just nanotechnology]. It would be a grave and possibly dangerous distortion if nanotechnology were to come to be seen as a discipline that raises unprecedented ethical and moral issues. In this respect, I think it genuinely does differ from some aspects of biotechnological research, which broach entirely new moral questions".

On the other hand, Jean-Pierre Dupuy (2007; 2009a; 2009b) says that once nanoethics is possible and necessary, it must be developed confronting questions regarding moral philosophy, avoiding, then, common pitfalls already witnessed. He believes that nanotechnologies do pose new and unique issue to society and for so, nanoethics is really needed. With respect to nanotechnology, the questions of moral philosophy that must be raised are not new, but have to be discussed. These are: "the artificialization of nature; the question of limits; the role of religion; the finiteness of the human condition as something with a beginning and an end; the relationship between knowledge and know-how; the foundations of ethics" (Dupuy, 2007, p. 237). According to Dupuy (2007), this is the return of the ever present theme of whether mankind is -or will ever be- able to cope with technological progress that breaks social paradigms and fundamentally alters the very fabric of human nature. More than that: reflections are needed upon how to restrain the development and use of nanotechnology in ethically and responsible manner.

Society must be ready for the unknown. It is imperative. Although the world is unlikely to be ruled by self-replicating machines, unintended; cumulative; chronic and other unforeseeable consequences can be expected, as new technologies develop.

For being disruptive and transformative, nanotechnology outcomes cannot be anticipated. However, to monitor its development may help society reap the benefits and minimize the adverse outcomes (Litton, 2007).

The above paragraphs were just a reflection on such emerging research field. Whether or not nanotechnology needs its own ethics, or whether or not nanoethics is really a new branch of ethics are not questions that need or intend to be answered here. But, despite any point of view, a common opinion is that real-time ethical reflection on all aspects of nanoscience and nanotechnology should to begin now, and should include health and environmental impacts, human enhancement and individual privacy issues. Ideally, studies on ethics *from* or *for* nanotechnology should experience the same exponential growth that is taking place in nanotechnology.

CONCLUSIONS

Whenever new technologies emerge, people project all sorts of fears and hopes in it. Expectations, such as more jobs, exciting economic growth and opportunities, more leisure, better education, or enhanced democracy and public debate, come up.

Nanotechnology, as any other technology, can be presented as human salvation that will solve all problems, or damnation, as the major source of problems at the present epoch. It encompasses an extraordinary diversity of technological approaches currently under development. Due to its innovative characteristics, nanotechnology can take society into a novel field of cultural experiences, where the very nature of human identity and social relations will be profoundly changed. Some authors believe that nanotechnological revolution will radically reshape economies, labor relations, social structures, and also the relationship with the natural world.

New and emerging technologies imply in new and emerging risks. In such context, nanotechnology can be considered an important source of known and unknown risks, due to the increasing uncertainties posed by this technology, regarding unique challenges in several aspects, such as occupational health and safety, environmental, legal, economic political, social end ethical.

Despite the importance of aspects described above, governments and industry are dramatically investing in nanotechnology research and product development. Very little is being invested on the science of impacts, the so-called uncertainties or scientific unknowns. So, produced non-knowledge is not being properly addressed as a consequence of lower financial resources supporting such activity.

The development of assessment methodologies are not in line with the pace at which marketing of nanoproducts become reality. Health & Safety and also environmental impacts remain poorly studied. The manipulation and deliberate manufacture of nanomaterials that are different from anything that occurs in nature, can lead to unintended and even unknown consequences to the environment and health and safety of workers. They can be exposed to new hazards in manufacturing, maintenance, decommissioning and recycling of nanomaterials-containing goods. A detailed investigation of nano-enhanced products is needed at all stages of the product life cycle to avoid both environmental and health impacts.

As being innovative and disruptive, nanotechnology can lead to changes in the manufacturing industry profile, resulting in organizational changes, with new division of labor and new competence demands to new job positions. Professional qualification will be dramatically impacted due to new skills and knowledge required by the market. Because a great deal of previous information will become useless, workers will have to be re-qualified to still integrate the workforce.

Along with the adoption of new technologies in capitalist economies, workers have tried to gather together in order to defend common interests, such as, minimize losses and the uneven distribution of profits, ensuring that the benefits arising from technological development are reaped by everyone. They claim for agreements which include clauses about the rights to information regarding new technologies and their applications in the workplace, environmental protection and workers health; employers' duties in the prevention of nanotechnologies impacts on workers; and personnel requalification.

Innovation in the field of nanotechnology is far ahead of the policy and regulatory issues. The deficits include, for instance, lack of programs for monitoring and controlling the impacts of nanomaterials in the workplace; lack of harmonization of risk assessment and management procedures; and fragmented regulatory structures, both nationally and internationally. This large gap of knowledge, regarding nanomaterial impacts, may hamper the ability to ensure workers' and consumer's safety as well as effectively regulate the production of nanomaterials. Resulting costs are, thus, transferred to society as a whole in several ways: workers are exposed to unknown hazards that may jeopardize health; population in general can be exposed to the same hazards if the surrounding environment is polluted by nanoparticles released by nanoproducts facilities; and at last, nanoparticles leach from products to accumulate in the environment.

Nanotechnologically improved goods are being produced and commercialized; workers and consumers are being exposed to the unknown potential risks of nanotechnology and mostly uninformed about them. There is no information processess driven to society as a whole or to workers specifically in order to raise awareness about nanotechnology consequences. The lack of public debate, social control and participation in decision-making processes may lead to an uneven distribution of knowledge. This, in turn, can create a nanodivider between those who can benefit from the innovations and those who can neither afford nor access them. Poor countries will become poorer and consequently more likely to be increasingly dependent on richer and technologically developed countries. They will not probably groove the benefits from environmentally and friendly nanotechnological solutions, products and processes. Instead, cheap and outdated technology will be exported to them, as suitable and affordable solutions according to their current status on global economy.

According to the scenario described above, the *Capital* itself is the only one which benefits, assuring its production and reproduction, regardless of the negative consequences towards society, which are: the worsening occupational safety and health conditions and working environment and uneven distribution of risk uncertainties. The private appropriation of public resources aiming at producing knowledge to sustain the capital reproduction leads to an uneven and unfair distribution of risks and uncertainties throughout the society.

To avoid uneven and unfair distribution of positive and negative outcomes, both short- and long-term impacts from nanotechnology must be addressed, including ethical and social ones, intended and unintended consequences and effects of its use. It must occur at each stage of the development process. More robust strategies and policies can be developed, if uncertainties are better studied and understood. Public and private sectors must invest in it too. It is the only way to keep occupational health and safety matters in pace with nanotechnological innovation.

More than this, a joint effort at local, national and international levels is necessary to promote a broad discussion of what nanotechnology really is, what it brings and what it could do, their positive and negative uses and consequences, whether or not nanotechnology enhancements are beneficial to society (workers included), as well as what can be done to promote a positive outcome and prevent harmful ones.

After all if nanotechnology, or any other technology, itself is subject to societies' purposes, it can be constructed and reconstructed in line with societies projects and values, thus, contributing to societies' development.

ACKNOWLEDGEMENTS

None declared.

CONFLICT OF INTEREST

The authors confirm that this chapter content has no conflict of interest.

REFERENCES

Allhoff, F., & Lin, P. (2006). What's so special about nanotechnology and nanoethics? *International Journal of Applied Philosophy, 20*(2), 179-190. Retrieved from http://files. allhoff.org/research/Special_Nanotechnology_Nanoethics.pdf

Allhoff, F., & Lin, P. (2007). Nanoscience and nanoethics: Defining the disciplines. In F. Alhoff, P. Lin, J. Moor, & J. Weckert. (Eds.). *Nanoethics: The ethical and social implications of nanotechnology* (pp. 3-17). New Jersey: John Wiley & Sons.

Arcuri, A. S. A., Grossi, M. G. L., Pinto, V. R. S., Rinaldi, A., Pinto, A. C., Martins, P. R., & Maia, P. A. (2009). Developing strategies in Brazil to manage the emerging nanotechnology and its associated risks. In I. Linkov, & J. Steevens (Eds.). *Nanomaterials: Risks and benefits* (pp. 299-307). Dordrecht: Springer.

Australian Council of Trade Union Fact Sheet (2009). *Nanotechnology: Why unions are concerned.* Retrieved from http://www.actu.org.au/Images/Dynamic/attachments/6494/actu_factsheet_ohs_-nanotech_090409.pdf

Australian Manufacturing Workers' Union. (2008). *Inquiry into nanotechnology in New South Wales.* Retrieved from https://www.parliament.nsw.gov.au/prod/parlment/committee.nsf/0/d2be239f1cf0c161ca257426007f3356/$FILE/AMWU%20Submission%20to%20Inquiry%20into%20nanotechnology%20in%20New%20South%20Wales.doc

Bainbridge, W. S. (Ed.). (2007). *Nanotechnology: Societal implications.* (Vols. 1-2). Netherlands: Springer.

Ball, P. (2003). *Nanotechnology in the firing line.* Retrieved from http://nanotechweb.org/cws/article/indepth/18804

Barry, B. E. (2008). The state of the science: Human health, toxicology and nanotechnological risk. In J. A. Shatkin (Ed.). *Nanotechnology health and environmental risks.* New York: CRC Press.

Bensaude-Vincent, B. (2013). Decentring nanoethics toward objects. *Etica & Politica, XV*(1). 310–320. Retrieved from http://www.openstarts.units.it/dspace/bitstream/10077/8903/1/BENSAUDEVINCENT.pdf

Borm, P. J., & Kreyling, W. (2004, May). Toxicological hazards of inhaled nanoparticles: Potential implications for drug delivery. *Journal of Nanoscience and Nanotechnology, 4*(5), 521-531.

Boschen, S., Kastenhofer, K., Marschall, L., Rust, I., Soentgen, J., & Wehling, P. (2006, December). Scientific cultures of non-knowledge in the controversy over genetically modified organisms: The cases of molecular biology and ecology. *GAIA: ecological perspectives for science and society, 15*(4). 294-301. Retrieved from http://docserver.ingentaconnect.com/deliver/connect/oekom/09405550/v15n4/s12.pdf?expires=1401635287&id=78312587&titleid=6690&accname=Guest+User&checksum=AFB77ED81ED402CF1F186C99EDDF35D9

Bostrom, N. (2002). Existential risks: Analyzing human extinction scenarios and related hazards. *Journal of Evolution and Technology, 9*, 1-45. Retrieved from http://www.nickbostrom.com/existential/risks.pdf

Brown, C., & Campbell, B. (2005). *The impact of technological change on work and wages.* Retrieved from http://www.irle.berkeley.edu/worktech/worktech.pdf

CEFET-SP. (2000). Tecnologia e microeletrônica. Retrieved from http://www.cefetsp.br/edu/eso/globalizacao/textocut.html

Chan-Remillard, S., Kapustka, L., & Goudey, S. (2009). Nanotechnology: The occupational health and safety concerns. In I. Linkov, & J. Steevens (Eds.). *Nanomaterials: Risks and benefits* (pp. 53-66). Dordrecht: Springer.

De Jong, W. H., & Borm, P. J. A. (2008). Drug delivery and nanoparticles: Applications and hazards. *International Journal of Nanomedicine, 3*(2). 133-149. Retrieved from http://www.ncbi.nlm.nih.gov/pmc/articles/PMC2527668/pdf/ijn-0302-133.pdf

Del Castillo, A. M. P. (2010). *The EU approach to regulating nanotechnology.* Brussels: ETUI.

Departamento Intersindical de Estatística e Estudos Socioeconômicos. (2008). *Nanotecnologias e os trabalhadores do ramo químico: Um registro da trajetória dos químicos da CUT no Estado de São Paulo.* Retrieved from http://www.nano.iiep.org.br/node/782

Dupuy, J. P. (2007, May/June). Some pitfalls in the philosophical foundations of nanoethics. *Journal of Medicine and Philosophy, 32*(3), 237–261.

Dupuy, J. P. (2009a). Technology and metaphysics. In J. K. B. Olsen, S. A. Pedersen, & V. F. Hendricks, (Eds.). *A companion to the philosophy of technology* (Chap. 38, pp. 214-217). West Sussex: Willey-Blackwell.

Dupuy, J. P. (2009b). The critique of the precautionary principle and the possibility for an enlightened doomsaying. In J. K. B. Olsen, S. A. Pedersen, & V. F. Hendricks (Eds.). *A companion to the philosophy of technology* (Chap. 37, pp. 210-213). West Sussex: Willey-Blackwell.

Dupuy, J. P. (2009c). *On the origins of cognitive science: The mechanization of the mind.* (M. B. De Bevoise, Trad.). Massachussets: The MIT Press.

ETC Group. (2005). *The potential impacts of nano-scale technologies on commodity markets: The implications for commodity dependent developing countries.* Retrieved from http://www.etcgroup.org/sites/www.etcgroup.org/files/publication/45/01/southcentre.comm odities.pdf

European Agency for Safety and Health at Work. (2012). *Risk perception and risk communication with regard to nanomaterials in the workplace.* Luxembourg: Publications Office of the European Union. Retrieved from https://osha.europa.eu/en/publications/literature_reviews/ risk-perception-and-risk-communication-with-regard-to-nanomaterials-in-the-workplace

European Agency for Safety and Health at Work. (2013a). *Green jobs and occupational safety and health: Foresight on new and emerging risks associated with new technologies by 2020.* Luxembourg: Publications Office of the European Union. Retrieved from https://osha. europa.eu/en/publications/reports/green-jobs-foresight-new-emerging-risks-technologies

European Agency for Safety and Health at Work E-Facts (2013b). *E-fact 74: Nanomaterials in maintenance work: Occupational risks and prevention.* Retrieved from https://osha.europa. eu/en/publications/e-facts/e-fact-74-nanomaterials-in-maintenance-work-occupational-risks-and-prevention.pdf

European Environment Agency. (2001). *Late lessons from early warnings: The precautionary principle 1896-2000.* Luxembourg: Publications Office of the European Union. Retrieved from http://www.eea.europa.eu/publications/environmental_issue_report_2001_22

European Trade Union Confederation. (2008a). *ETUC resolution on nanotechnologies and nanomaterials.* Retrieved from http://www.etuc.org/sites/www.etuc.org/files/ETUC_ resolution_on_nano_-_EN_-_25_June_08_2.pdf

European Trade Union Confederation. (2008b). *ETUC wants precautionary approach applied to nanotechnologies.* Retrieved from http://www.etuc.org/a/5159

Ferrari, A., & Nordmann, A. (2010, August). Beyond conversation: some lessons for nanoethics. *NanoEthics, 4*(2), 171–181.

Flynn, R., Bellaby, P., & Ricci, M. (2006, January). Risk perception of an emergent technology: the case of hydrogen energy. *Forum: Qualitative social research, 7*(1). Article 19.

Foster, J. B. (2005, March). The treadmill of accumulation: schnaiberg's environment and marxian political economy. *Organization & Environment, 18*(1), 7–18. Retieved from http:// sociology.uoregon.edu/faculty/foster/Organization%20Environment-2005-Foster-7-18.pdf

Franssen, M., Lokhorst, G. J., & Van de Poel, I. (2009, Fall). Philosophy of technology. In E. N. Zalta, (Ed.). *The Stanford Encyclopedia of Philosophy.* Retrieved from http://plato.stanford. edu/archives/fall2009/entries/technology/#DevEthTec

Gonik, C. Y., DeCarlo, N., Turner, T., & Thomas-Muller, C. (2010, September/October). Finding the movement in the second contradiction. *Canadian Dimension, 44*(5). Retrieved from http://canadiandimension.com/articles/3267

Hanna, R. (2005, August). Kant and non conceptual content. *European Journal of Philosophy, 13*(2), 247–290.

Hannah, W., & Thompson, P. B. (2008, March). Nanotechnology, risk and the environment: a review. *Journal of Environmental Monitoring, 10*(3), 291-300.

Health and Safety Executive. *What is nanotechnology?* Retrieved from http://www.hse.gov.uk/nanotechnology/what.htm#potential-health-concerns

Intercâmbio, Informações, Estudos e Pesquisas. (2008). *Posicionamento sindical diante dos impactos éticos, sociais e ambientais da introdução de nanotecnologias nos alimentos, produtos e processos produtivos.* Retrieved from http://www.nano.iiep.org.br/sites/default/files/cartilha_posicionamento.pdf

International Risk Governance Council. (2007). *Nanotechnology risk governance: Recommendations for a global, coordinated approach to the governance of potential risks.* Geneva: Author.

International Risk Governance Council. (2010). *The emergence of risks: Contributing factors.* Geneva: Author.

Invernizzi, N., & Foladori, G. (2005). Nanotechnology and the developing world: Will nanotechnology overcome poverty or widen disparities? *Nanotechnology Law & Business Journal, 2*(3). Article 11. Retrieved from http://estudiosdeldesarrollo.net/administracion/docentes/documentos_personales/11947LBJ.pdf

Keiper, A. (2007, spring). Nanoethics as discipline? *The New Atlantis: A Journal of Technology & Society, 16,* 55-68. Retrieved from http://www.thenewatlantis.com/docLib/TNA16-Keiper.pdf

Kreyling, W. G., Semmler-Behnke, M., & Möller, W. (2006a, October). Health implications of nanoparticles. *Journal of Nanoparticle Research, 8*(5), 543-562.

Kreyling, W. G., Semmler-Behnke, M., & Möller, W. (2006b, spring).Ultrafine particle–lung interactions: does size matter? *Journal of Aerosol Medicine, 19*(1), 74-83.

Kurzweil, R. (2005). *The singularity is near: When humans transcend biology.* New York: Viking.

Linkov, I., & Satterstrom, K. (2008). Nanomaterial risk assessment and risk management: review of regulatory frameworks. In I. Linkov, E. Ferguson, & V. Magar (Eds.). *Real time and deliberative decision making: application to risk assessment for non-stressors* (pp. 129-158). Amsterdam: Springer.

Litton, P. (2007, January/February). Nanoethics?: What's new? *Hastings Center Report, 37*(1), 22-25. Retrieved from http://papers.ssrn.com/sol3/papers.cfm?abstract_id=1311533

Lowry, G. S., & Casman, E. A. (2009). Nanomaterial transport, transformation and a fate in the environment. In I. Linkov, & Steevens, J. (Eds.). *Nanomaterials: Risks and Benefits,* Dordrecht: Springer.

Mansoori, G. A., & Soelaiman, T. A. F. (2005, June). Nanotechnology: an introduction for the standards community. *Journal of ASTM International, 2*(6), 1-21. Retrieved from http://www.uic.edu/labs/trl/1.OnlineMaterials/nano.publications/05J.ASTM.Intl.June05.Vol2.No6.pdf

Nanoaction Project (2007). *Principles for the oversight of nanotechnologies and nanomaterials.* Retrieved from http://www.centerforfoodsafety.org/files/final-pdf-principles-for-oversight-of-nanotechnologies_80684.pdf.

National Science Foundation (2000). Nanotechnology definition (NSET, February 2000). Retrieved from http://www.nsf.gov/crssprgm/nano/reports/omb_nifty50.jsp.

National Science Foundation (2001). *Societal implications of nanoscience and nanotechnology.* Arlington: Author. Retrieved from http://www.wtec.org/loyola/nano/NSET.Societal. Implications/nanosi.pdf.

Nurock, V. (2010). Nanoethics: Ethics for, from or with nanotechnology. *International Journal for Philosophy of Chemistry, 16,* 31-42. Retrieved from http://www.hyle.org/journal/issues/16-1/nurock.pdf

O'Connor, J. (n.d.). *Selling nature.* Retrieved from http://www.sagepub.com/upm-data/13298_Chapter_9_Web_Byte_James_O'Connor.pdf

O'Connor, J. (1988). Capitalism, nature, socialism: a theoretical introduction. *Capitalism, Nature, Socialism, 1*(1), 11-38. Retrieved from http://www.vedegylet.hu/okopolitika/O'Connor%20-%20Capitalism,%20Nature,%20Socialim.pdf

O'Connor, J. (1996). The second contradiction of capitalism.In Benton, T. (Ed.). *The greening of Marxism.* New York: The Guilford Press.

O'Connor, J. (1998). *Natural causes: Essays in ecological Marxism.* New York: Guilford.

Papadopoulos, G., Georgiadou, P., Papazoglou, C., & Michaliou, K. (2010, October). Occupational and public health and safety in a changing work environment: an integrated approach for risk assessment and prevention. *Safety Science, 48*(8), 943-949.

Ribeiro, M. G., & Pedreira Filho, W. R. (2006). Risk assessment of chemicals in foundries: the international chemical toolkit pilot-project. *Journal of Hazardous Materials*, A136.432-437.

Rice, J. (2009, June/August). The transnational organization of production and uneven environmental degradation and change in the world economy. *International Journal of Comparative Sociology, 50*(3-4), 215–236.

Rifkin, J. (1995). *The end of work: The decline of the global labor force and the dawn of the post-market era.* New York: Putnam.

Roco, M., & Bainbridge, W. (Eds.). (2003). *Converging technologies for improving human performances: nanotechnology, biotechnology, information technology and cognitive science.* Retrieved from http://www.wtec.org/ConvergingTechnologies/Report/NBIC_report.pdf.

Schnaiberg, A. (1997). Sustainable development and the treadmill of production. In Baker, S. (Ed.). *The politics of sustainable development: Theory, policy and practice within the European Union.* London: Routledge Press.

Slovic, P. (2000). *The perception of risk.* London: Earthscan.

Spence, M. (2000, autumn). Capital against nature: James O'Connor's theory of the second contradiction of capitalism. *Capital & Class, 24*(3), 81-110.

Walker, M. (2014). Big and technological unemployment: Chicken little versus the economists. *Journal of Evolution and Technology 14*(1), 5-25. Retrieved from http://jetpress.org/v24/walker.pdf

Warheit, D. B., Reed, K. L., & Sayes, C. M. (2009, July). A role for nanoparticle surface reactivity in facilitating pulmonary toxicity and development of a base set of hazard assays as a component of nanoparticle risk management. *Inhalation Toxicology, 21*(supplement 1), 61-67.

Weekly, D. J. (2008). *The commercialization of nanotechnology.* Retrieved from http://uwyo.coalliance.org/fedora/repository/wyu:114/Nanotechnology_Word.pdf

Wynne, B. (2001). *Managing and communicating scientific uncertainty in public policy.* Paper presented at the Harvard University Conference on Biotechnology and Global Governance: Crisis and Opportunity. Kennedy School of Government, Cambridge.Retrieved from https://googledrive.com/host/0B1arYAWdEZhkbnJjcEt2NlY4RTg/HARVARD%20UNCS %20PAPER.pdf

Yah, C. S., Simate, G. S., & Iyuke, S. E. (2012, April). Nanoparticles toxicity and their routes of exposures. *Pakistan Journal of Pharmaceutical Sciences, 25*(2), 477-491.

CHAPTER 3

Occupational Health and Safety Knowledge and Practice

Marcela G. Ribeiro[*] and Fernanda F. Ventura

Fundação Jorge Duprat Figueiredo de Segurança e Medicina do Trabalho – Fundacentro, São Paulo, Brazil

Abstract: Changes in the world of work have been witnessed worldwide. They have profoundly affected the working life, the occupational health and safety professions structure, and occupational health and safety professional role. This chapter invites to reflections concerning needs on occupational health and safety professional competencies demanded by the changing world of work. Occupational health and safety professionals will be a necessary part of doing business worldwide, as researchers, educators, or practitioners. Nevertheless, competencies these professionals must acquire are at stake. Training and education must be redesigned for creating competent critical mass of occupational health and safety experts, able to understand and to deal with emerging issues in the changing world of work. Multiple disciplinary learning approaches are needed to tackle today's complex problems of working environments. The occupational health and safety professionals must find balance between *knowledge breadth* and *depth*. Changing values at university is an opportunity to enhance the value of occupational health and safety for a wider audience in an environment where occupational health and safety issues are usually under evaluated. Beyond technical expertise, skills are needed to move forward. Resilience is certainly the most important to be developed. To be resilient, professionals must be self-reliant, requiring a continuous openness towards life-long learning. The occupational health and safety professional must now know how to learn during an entire lifespan. Despite theoretical knowledge gathered from studies already conducted on several aspects, occupational health and safety conditions worsen. Attempts should be made to strengthen knowledge-practice link. Researchers, professors, students and practitioners must face this biggest challenge, which can determine the success or the failure on the enhancement of occupational health and safety conditions.

Keywords: Breadth of knowledge, changing world of work, commitment, communication, continuing learning, depth of knowledge, life-long learning, long-lasting research, multiple disciplinary approach, OHS profession,

*Corresponding author Marcela G. Ribeiro: Fundação Jorge Duprat Figueiredo de Segurança e Medicina do Trabalho, Rua Capote Valente, 710 CEP: 05409-002, São Paulo, SP, Brazil; Tel: +55 11 3066-6075; Fax: +55 11 3066-6341; E-mail: marcela.ribeiro@fundacentro.gov.br

Marcela G. Ribeiro (Ed)
All rights reserved-© 2014 Bentham Science Publishers

OHS professional skills, OHS research, research and practice gap, resilience, self-reliance, teamwork, training needs, university education.

INTRODUCTION

The world of work is facing transformations never seen before, with the introduction of new technologies and processes. This is coming along with changes in the traditional characteristics of workplaces and workforce and the introduction of new working demands, modifying work relations and organization. In this context, there is a need for improved knowledge; education; training and skills in occupational health and safety (OHS), in order to pace with the new risks and health threats arising from the changes.

Herrera (2012, p. 951) quotes the report *New trends in accident prevention due to the changing world of work*, published by the European Agency for Safety and Health at Work (2002b), to list the main changes affecting the working life, which can be, among many others:

> "(1) changing industrial organizations, (2) the free market, privatization, downsizing, subcontracting, (3) adoption of new technologies, (4) the growing use of remote operations, homework, changes in working hours, work pace and workload, (5) changing labor market with an increase in part-time jobs, temporary work, self-employment, women in employment, the aging of the workforce".

Such changing work structures bring to the spotlight risks not faced before, as new types of hazards and changes in the nature of accidents. Papadopoulos, Georgiadou, Papazoglou & Michaliou, (2010, p. 947) argue that "risk assessment and prevention in the changing work environment is a complex multi-criteria decision-making problem. Appropriate methodologies are required, as well as long lasting research regarding many parameters to reach an effective confrontation of the new difficulties arising in the field of occupational safety & health". In this context, formerly functional recipes and responses are becoming outdated (Wilpert, 2009).

The global economic, political, and cultural changes modified the workplace and have also profoundly affected the structure of OHS professions. These continuous and rapid changes should be approached by a dynamic OHS management, intervention and preventive systems, emphasizing *e.g*, participation, leading performance, communication and life-long learning, which, in turn, require a dynamic OHS professional.

To reach this goal, some questions must be answered. Among them, how to provide an OHS professional with suitable qualities and trained for this new work environment? What are the changing demands that arise from such globalized workplace and work conditions, in terms of professional OHS qualification? Can these changes in the work environment be effectively approached, only changing methods and tools? If not, what are the demands and where are they ultimately needed in order to fulfill these new OHS needs? This chapter is not intended to give clear and finite answers to the questions proposed. Rather, it is intended to invite the reader to reflections concerning the education and qualification needs, demanded by the changes in the world of work. The authors intend to bring some important questions to light and give birth to discussions that contribute to an OHS professional, directed and formed to the work environment where this professional is needed.

With this in mind, the chapter begins with a general overview on different aspects of the changing world of work and its implications on OHS related issues. The challenges faced by the OHS professionals and some education and training needs in the OHS field will be presented. Additionally, some of the qualifications and characteristics that OHS professionals should have in order to attend these new OHS demands will be summarized. To finish, the gap between knowledge and practice will be approached and all aspects discussed through the text will be summarized and possible conclusions presented.

OVERVIEW

The world of work has been changing fast due to the development and adoption of new technologies, as well as to respond to business demands (European Agency for Safety and Health at Work [EU-OSHA], 2002d). In the last few years, researchers have pointed the need to manage safety and health consequences of the changing structures of employment that followed the development of what is referred as the post-modern society (EU-OSHA, 1998; 2002c)

In the so-called *'Changing World of Work'* (EU-OSHA, 1998; 2000), the traditional concept of work changed from the one with a long lasting contract and a formal work site to where the worker went every workday. New forms of work arose, such as work at home, telework, self-employment, subcontracting, outsourcing work, temporary work and employment in micro and small firm, among many others. A general trend towards flexible and irregular working hours is observed. Although it can be seen at first as a good work change, such changes usually increase work pace, workload, work hours, accident rate, as well as stress. Demographic changes have shown an ever increasing number (rapidly growing) of older workers and of working women. It is also observed an increasing in the number of atypical workers (part-time, temporary and self-employed) that usually fall outside the reigning OHS systems (and policies), also being set aside from the internal safety organization of the company to which they are connected at the time (EU-OSHA, 2002c).

The brief overview on the emerging trends and possible OHS implications, presented on the following paragraphs, is based on the EU-OSHA reports *Research on Changing World of Work* (2002c) and *The changing world of work. Trends and implications for occupational safety and health in the European Union* (2002d).

Changes, as flattering management structure or fragmentation of traditional larger enterprises, have brought management and structural transformations to these now smaller organizations, which affect the health and safety management and can result, as pointed out by EU-OSHA report (2002d), in: reduced central OHS management control; lack of clarity in OHS responsibilities; delegation of OHS responsibilities to line managers, which can have, in turn, limited time, resources and qualification; reduced direct employment of safety and health personnel and reduced capacity to deal with restrictions on OHS budgets. In this context, organizations may no longer employ health and safety specialists, either internally, or through external preventive services. As a consequence, its capacity to respond and deal with OHS issues is dramatically reduced. Previous studies indicated, for example, an increase in stress-related and musculoskeletal disorders among those who survived to such structural and organizational changes (EU-OSHA 2002d).

In all sectors, a growing need for information and communication is observed. Actually, information technology has had a profound effect on the nature of some tasks and on the work organization. It has strongly reduced the distance between workers and employers and is currently reducing actual national frontiers. Computers have changed the way people work, under different aspects. Information technology, e-business and virtual company workers are highly mentally demanded. The constant use of display screen equipment commonly leads to musculoskeletal disorder risks and eye fatigue among workers from these areas. Interfaces and equipment are not always friendly, and proper training may not be easily available (EU-OSHA, 2002b; 2002c; 2002d).

The growing of the service sector, to a point that a never ever seen proportion of workers are now in this area, is another remarkable change. The increasing employment in the service sector leaded to an increase in the number of small and medium enterprises (SME), with temporary workers. Such sector usually does not have well developed and established OHS systems, nor traditions. Additionally, this increase in SME and micro-business and their lack of an 'OHS culture' usually leads to higher accident rates, attributed to a lesser presence of formal management structures, resources and awareness concerning the importance of OHS related issues. Such enterprises, usually, do not realize that health and safety at work is more than a legal requirement (EU-OSHA, 2002b; 2002c; 2002d).

Changes in work organization include, for example, teleworking, which can result in possible social isolation and ergonomic problems, since home is not usually designed as workplaces. According to Boedeker & Klindworth (2007), constant work reorganization was identified as a contributing factor to some major accidents and can lead to an increase in stress and fatigue levels. Additionally, in non-traditional workplaces (e.g. home-offices), confusion over OHS matters and responsibilities may arise.

Such changes also comprise increase of 24-hour working; increased working shifts and working hours; higher pace of work; higher intensity of work; and rapidly changing tasks and working patterns. In a general basis, more flexible working patterns mean more workers exposed to the potential adverse outcomes of both shift-work and night-shift work (EU-OSHA, 2002d). Increased work intensity means "extended practice of weekend work, the increase of working time schedules with irregular and less predictable working hours, and the use of both very limited hours (involuntary part-time work) and excessively long working hours (involuntary overtime)" (EU-OSHA, 2002d, p. 3). All these changing factors make risk assessment more complex and difficult.

Interpersonal relationships can also change as contractors and outsources from different origins may work together, or one is working for another, at a same workplace. This was pointed out as a source of uncertainties and misunderstandings concerning OHS responsibilities, such as, who should be responsible for training workers on OHS related matters; who should provide personal protective equipment; or yet, who is responsible for assess and manage occupational risks (EU-OSHA, 2002d).

Changes in employment relations still include self-employment; increased part-time work; temporary work; as well as any kind of precarious working relation. Workers in such situation may experience social isolation and stress as a consequence of related uncertainties. It is known that permanent workers are exposed to high-speed work and present increased stress levels and more mental health problems, such as depression (EU-OSHA, 2002c; Letourneux 1998; Benavides & Benach, 1999). On the other hand, "working conditions, in particular physical constraints and conditions of employment, of precarious workers are worse than those of permanent workers: more work in painful positions, more exposure to noise, more repetitive tasks and movements, less access to training, less autonomy over their work and time and less access to participation" (EU-OSHA, 2002c, p.6-7). Even though such workers are the most exposed to physical and chemical hazards, in less skilled and potentially monotonous positions, they may not be included on OHS programs and services, since the coverage by OHS systems is not always clear (EU-OSHA, 2002c; 2002d).

Changes in workforce include aging, higher percentage of women in paid employment and more immigrant workers. Older workers, for instance, need specialized assistance to deal with and to adapt to the increased and ever-changing job demands resulting from new organizational practices and new technologies adoption. They also take longer to gather and learn new knowledge, although the time needed for this purpose is barely available (EU-OSHA, 2002d). Regarding working women, despite their increased participation in labor market, most tasks and equipment have not been adapted to them yet. For those who still are responsible for taking care of their families, it becomes difficult to accomplish traditional working shifts. Besides that, differences regarding the type of jobs and tasks performed by men and women still remain, in addition to distinct employment contracts and opportunities for career development (EU-OSHA, 2002d). In addition, the EU-OSHA report (2002d) report shows that women usually have high demands at their job positions, despite their limited individual control over work. At last, as said by Sassen (1998, p.320), "the disruption of traditional work structures as a result of the introduction of modern modes of production has played a key role in transforming people into migrant workers", which are, in general, very poorly educated and qualified. Such workers usually experience language problems, which impacts their OHS training and information. They are still concentrated in unskilled jobs, with poor working conditions. All these issues impacts seriously on OHS management.

The European Agency for Safety and Health at Work states that "changes in management structures and responsibilities affect the management of occupational safety and health" (EU-OSHA, 2002d, p.2). Consequently, such changes affect the approaches needed to effectively improve health and safety in organizations. It is not easy for OHS professionals to deliver suitable information and support to these fragmented groups under constant changing environments. To cope with these challenges, and ensure health and safety at work, collaborative actions are necessary, as action research, intervention studies and learning networks. Cooperation, coordination, provision and sharing of qualitative and quantitative information on OHS issues, between occupational health and occupational safety experts, are crucial steps into a new multidisciplinary approach, which is needed to be developed; implemented and assessed.

To accomplish the emerging OHS issues, it is necessary an OHS research and practice that keep pace with the changes observed and listed above and also the ones that are yet to come. Stakeholders within the health and safety system (i.e. employers, employees, unions, inspection and regulation agencies), must all discuss and adapt their roles and tasks to this fast changing reality. It must be observed that the more professionals are trained regarding OHS (at least minimally), the better for them. In an ideal scenario, workers from all professions within a company should have minimum information and effective training on OHS principles.

Changes and developments in working life can have positive or negative impacts on OHS. Such changes should be an opportunity to improve working conditions, instead of impairing OHS improvements. Work should not compromise the safety, the health, and above all, the workers life. Self-realization and staying healthy are still pursuing workers life objectives.

As pointed out previously (EU-OSHA, 2002c), despite society have advanced from a technological point of view, this is not the case concerning OHS issues. New or improved OHS systems are necessary for a safe and healthy workplace. In such context, it must be highlighted that the OHS professional is required as an essential part of the process. Despite their central role, it is still not well understood to what degree the current OHS professionals are dealing, in practice, with these new OHS implications and to what degree they have access to information or are properly qualified to accomplish the posed demands.

It is known that the "changing world of work is requiring worker's skills to be adapted more and more quickly" (EU-OSHA, 2002c, p. 15), and so it is to OHS professionals. Health and safety experts need to alter their previous, traditional way of thinking and working. It means that the usual scarce OHS human resources must be better and continuously qualified. More human and financial resources should be directed to produce professionals prepared with both knowledge and skills for the diversity of challenges they will probably face in the workplaces.

EDUCATIONAL NEEDS

The previously approached changes in the world of work and the consequent depletion of working conditions (e.g. increase of occupational injuries and illnesses) clearly make difficult the implementation of workplace health and safety programs. At the same time, the continuing lack of suitable OHS conditions in most small and many larger workplaces suggest needs for: (i) more OHS professionals with broader vision of OHS related issues; (ii) a rethinking of the OHS professions structure, as well as (iii) a great need for education; training and practical research in OHS field (National Research Council [NRC], 2000; Sauter & Rosenstock, 2000).

There are several authors that seek to address the current needs on OHS professional training and education. A comprehensive analysis of this topic was published in 2000 by the USA National Research Council in the book *Safe Work in the 21ˢᵗ Century: Education and Training Needs for the Next Decade's Occupational Safety and Health Personnel*. It assessed the supply and demand for OHS professionals, also evaluating their knowledge and abilities. Such analysis additionally sought to identify human resources needs, skills, and curricula required for the dynamic changing world of work. In many aspects, the results outlined are still up to date and some of its points of view will be presented in the next paragraphs.

The NRC (2000) states that OHS have what may be called as four core professions: occupational safety, industrial hygiene, occupational medicine and occupational health nursing, as well as other closely related OHS disciplines, such as ergonomists and occupational health psychologists. Such professionals can be found in industry and industry-like settings; consulting firms; government regulatory agencies, educational and research institutions; and hospitals and clinics.

A broad definition of occupational safety generally includes "the control of hazards and the prevention of accidents not only to protect the workforce, but also to protect the general public and the environment" (NRC, 2000, p. 39). Therefore, the occupational safety professional deals with "the interaction between people and the physical, chemical, biological and psychological effects (acute or chronic) that can adversely affect their well-being" (NRC, 2000, p. 39). The industrial hygiene, on the other hand, was reported as a field of applied science that concerns the anticipation, recognition, evaluation, and control of environmental factors in the workplace, whose effects on workers can range from discomfort to chronic diseases. Occupational Medicine, in turn, was defined by the report as "the area of preventive medicine that focuses on the relationships among the health of workers, the ability to perform work, the arrangements of work, and the physical, chemical and social environments of the workplace" (NRC, 2000, p. 57). Finally, occupational health nursing was defined as the specialty practice focused on promotion, prevention, and restoration of health within the context of a safe and healthy work environment. It is focused on the reduction of health hazards and prevention of adverse health effects (injury and illness) by adopting, for example, occupational and environmental safety programs. As described in the referred report, although each of the traditional OHS core professions emphasizes different aspects, all of them are focused on identifying hazardous conditions; materials; and work practices in workplaces, aiming at eliminate or reduce risks.

Researchers and practitioners in the above mentioned fields must have in mind that both the work itself, as well as the environment in which it is performed can impact workers' health, positively or negatively. They must also recognize that work's nature or context can be arranged to not affect or impair workers' health and that OHS promotion will be reached when job placements are in line with workers' physical and psychological potentialities or limitations (NRC, 2000; American College of Occupational and Environmental Medicine, 2014). So, in order to implement effective health and safety programs, OHS professionals need to take into account the changing workforce, not only respecting, but also reflecting the social diversity posed by a variety of ages; ethnicities and cultures. OHS education and research would need to include, as a social role, knowledge about how to encompass the age-differentiated physical and cognitive abilities; how to interact disabilities and chronic diseases with workplace demands and how to develop and use communication skills. This last one may be very helpful skill, to reach minority workers; literate and semi-literate workers and also foreign workers (NRC, 2000).

The transforming workplaces were pointed out as a complicating factor to the efficiency of OHS programs. As a consequence, training and delivery of OHS need to be different from the traditional ones that have been currently relied on, once the workplaces now are very different from the former fixed-sites (NRC, 2000). As OHS professional should also focus on underserved workers and workplaces (and it is a special difficulty faced nowadays), the simple increase or modification of OHS professionals training would not be enough. New models and methods for systematically OHS programs implementation should be developed and explored, aiming at reach the full spectrum of workers and workplaces.

Occupational safety and health professionals need to be requalified to deal with changes in work organization, such as "globalization; technology; other work-design factors and organizational design innovations" (NRC, 2000, p. 125). Potential new hazards are currently emerging from new technologies and through the way work is being performed in more global, unbounded and virtual organizations. Both content of work and employment relationship are being reshaped as the number of small and flattered business increases. In addition to this, factors such as the growing dependence on technology and the intensified pace of work were also identified as challenges for OHS professionals (NRC, 2000).

Workers are being pressured to provide innovative services and products with exceptional quality, at competitive prices and short delivery times. As a consequence, they are being highly demanded for new skills and continuous learning, in addition to the expanded job scope; increased work pace; and changing workplaces. At the same time, "workers face uncertainty in employment relationships, heightened interaction with both customers and other workers, and more involvement with information and communication technologies" (NRC, 2000 p. 125). Todays' societal characteristics, such as increasing numbers of single parents, dual-career households and aged dependents, can be considered an additional challenge to workers at managing multiple and competing interests in their personal and professional lives, acting as additional sources of time conflicts and increasing the possibility of dysfunctions and distress at workplaces (NRC, 2000).

Occupational health & safety professionals must be able to recognize and understand the effects of the structural and contextual work conditions and relations on workers physical and mental health, which implies the ability to identify the influences of such work organization factors on "physical; cognitive and behavioral functioning, including stress-related conditions and their link to health, safety, and performance" at work (NRC, 2000, p. 7). Finally, OHS professionals need to develop basic abilities to conduct strategies focused on prevention and perform organizational interventions, using their expertise on work organization to manage workplace stress and well-being issues.

The transmission of qualified knowledge to the future practitioners responsible for workplace health and safety is essential to improve OHS conditions, as well as to promote a suitable management of new risks emerging from new working conditions as a consequence of the already stressed changes in the world of work (workforce, workplace and work organization transformations). This must be directed both to workers and OHS professionals, as they must be in the forefront of changes in workplace and foresee possible health impacting issues. To accomplish todays' and future needs, the required OHS professionals must be prepared for discover and deliver new and enhanced methods to include and take into account the reality of small and medium-sized companies, which are increasingly decentralized with large mobile workforces. To be tailored to up-to-date needs and successfully implement workplace health and safety programs, it is common ground that OHS professionals must be qualified in a multidisciplinary/ interdisciplinary environment, through a more comprehensive and alternative learning experiences (NRC, 2000).

From the academic point of view, the number of Ph.D. being awarded must increase in all OHS related areas. In spite of that, the already existing training Ph.D. programs must be not only maintained but extended. Doctoral-level OHS professionals are needed to develop research, to teach and also train future practitioners (NRC, 2000). As reported by NRC (2000, p. 156), "although few if any industries require safety professionals with doctorates, a critical mass of such individuals is necessary both for the conduct of critical research in injury prevention and for the continued viability of the academic programs that produce practicing safety professionals at the associate, bachelor's, and master's degree level". Notwithstanding a great amount of OHS professionals works satisfactorily well combining formal education (for example, baccalaureate or master's level), continuing education and practical experience. A continuous provision of such professionals clearly depends upon Ph.D. scientist-educators. A critical mass of Doctoral-level OHS professional is also necessary to attract research funding and students to be properly trained. Then, recruiting graduate students to arouse their interest on OHS issues is an important step to develop relevant research and educational activities in this area. The evolving demands to meet the new challenges related with the changing work and workplaces request a much broader perspective than that taken in the past. This includes more emphasis and research needs on such fields as epidemiology, toxicology, ergonomics, respiratory disease, dermatology, behavioral sciences, health care cost control, and management. Specializing in a wide variety of fields is a very important way to foster the application of them to solve practical OHS problems (NRC, 2000).

According to EU-OSHA (2002d) published survey, research also needs to include studies regarding organizational interventions and targeted health effects, taking into account the new forms of work organization. It is also important to comprise the development and analysis of methodologies for risk assessment, along with studies about combination of risk factors and their application at workplaces. Other relevant research actions may include the evaluation of intervention studies focused on issues such as occupational stress and ergonomics at workplaces, aiming to prevent psychological and musculoskeletal disorders; the assessment of emerging examples of interventions, such as matching work to older workers, as well as analyses of already implemented interventions, which could help to determine success and failure factors; health studies and standardized research methodologies in work organizations and improved mechanisms for surveillance of changing work organization and its effects on job characteristics.

The USA National Research Council (NRC, 2000) pointed out that education at the baccalaureate, master and extension levels in all core OHS disciplines have their curricula mostly focused on technical and scientific aspects. Whereas safety and industrial hygiene are accentuated in engineering and physical sciences, biological; health-related and programmatic topics are the main focus in nursing and medicine. According to them, the lack of research and qualification in areas of increasing importance, such as behavioral health sciences (psychology, psychiatry, and social work), new forms of work organization, workforce ethnic and gender diversity, information systems, prevention and intervention, ergonomics, evaluation and assessment methods, communication and management must be addressed in order to respond to new dynamic workplace demands, and also as a pre-requisite for qualified professionals, which must be able to perform interventions in the workplace. Graduate training support on such areas must be broadened as well.

Taking into account the OHS requirements to deal with the changes in the world of work, additional topics in need of attention should be highlighted. Among them: methods for effective training adults and older workers; physical and psychological susceptibilities of workforce according to age, gender; health promotion and disease prevention; community and environmental concerns; and the ethical and societal implications of technological development, such as biotechnology and nanotechnology (NRC, 2000).

Despite stressing its enormous importance, responsibilities related with training and education of workers is not usually seen as a priority of OHS professionals. As a consequence, the quality and practical effects of such training and education programs is barely known. In fact, most graduates of OHS programs are admittedly ill prepared for this task. It was stated that a large number of graduate students have good technical knowledge, but are not skilled enough to provide a satisfactory adult education, training, and program evaluation. The final result is a very uneven training quality (NRC, 2000).

With this in mind and as already said before, the OHS practitioners that must deal and assist workers in this new world of work must go beyond the traditional training, which is no longer sufficient to accomplish adequate and necessary care of workers. To accomplish this task, other focus must be given to OHS professionals, which must be prepared to look for and open-minded to explore new models for implementing suitable OHS programs (NRC, 2000).

Among innumerous aspects to be considered in the training and education of the professionals to meet the new quests yet to be faced, taken into account the diversity in workforce and workplaces, the EU-OSHA (2002d) suggested that emphasis should be given on topics to promote the integration of work organization to occupational health into the curriculum of OHS specialists. Also, information concerning OHS topics should be given when training professionals from other disciplines, such as architects and designers.

It is noteworthy to mention that, in the presented context, an essential aspect of training must be on prevention. Prevention should start as earlier as possible; awareness activities to improve safety culture should be started at school level. The inculcation of a culture of safety and health in the general public, according to the USA National Research Council (NRC, 2000), is a broad, long-term, multifaceted project that requires leadership and that need to encompass methods to disseminate information (e.g., mass media vehicles, such as Internet and school education, among other strategies), intending to reach the society as a whole (youth, parents, workers and employers).

In this respect, it was suggested that government agencies, unions, industries and employers should work together in order to learn from successfully established OHS programs. Such experience could be useful to strengthen new workers training programs. In the same vein, funding and grants from governmental agencies should be given to research and extension projects aimed at developing new OHS training tools and models that could reach a broader scope of workers, such as those from small and service-sector companies. This is substantial, since in such businesses, managers usually have a lot of responsibilities and little (if any) formal education or training in OHS. The use of "new learning technologies, the development of a recommended set of basic competencies, and the creation or recognition of a new category of OHS personnel: the occupational safety and health manager" is strongly encouraged by NRC (2000, p. 194).

It is also important to provide a continuing education in order to promote lifelong learning. According to the Encyclopaedia of Occupational Health and Safety (Stellman 1998), continuing education can be formal or informal, voluntary or mandatory. It can be performed via, for example, conferences; workshops; lectures; journals clubs and seminars. The result from a continuing OHS education throughout working life is a lifelong learning, which is defined by the

Commission of the European Communities (2001, p. 33) as "all learning activity undertaken throughout life, with the aim of improving knowledge, skills and competence, within a personal, civic, social and/or employment-related perspective". Thus, continuing education is an essential tool for keeping OHS professionals up-to-date, who by the lifelong learning will in turn be more prepared to face current and future changes at work.

The USA National Research Council (NRC, 2000) still addresses the multidisciplinary learning as another important aspect to provide the knowledge, and enhance skills and abilities required to good OHS practice at workplaces, such as problem solving capability.

From the educational point of view, multidisciplinary approaches can be introduced by traditional education or by other alternative programs (such as the distance learning, which has a comparative broader range of public). But at the same time, as stressed by the USA National Research Council, simply adding more courses (and disciplines) to the curriculum of OHS core areas is not feasible, because it could add months or years to any of the programs - which would probably have a strong negative effect on the attractiveness of the programs to prospective students. Substitutions for areas currently in the curriculum will also be difficult due to previously established programs.

As seen in the paragraph above, curricula, alternative learning methods and new topics to be embedded on graduate courses are, with no doubt, important aspects to bring up reflections. Nevertheless, the way knowledge is taught to the OHS professionals must also be addressed. It will reflect directly on the way OHS professionals see and understand their role on the workplace, and also how they respond to the workplace today's needs.

In order to positively impact health and safety practice and research, OHS issues must be approached in an integrated and transversal manner among curricula disciplines. According to Limborg (2001, p. 170), "interdisciplinary cooperation has shown itself to be a prerequisite for being able to analyze and develop solutions to the often very aggregated and complex problems characterizing today's working environment". He further states that these intricate questions are unlikely to be solved by a single professional from only one discipline.

Rosen, Caravanos, Milek & Udasin (2011), in turn, believe that the integration of the OHS four core disciplines and the other closely-related is important for future practitioners to develop a comprehensive view of workplace injuries and illnesses causes, consequences and ways of prevention. In the paper *The professional working environment consultant – A new actor in the health and safety arena*, the authors describe how an interdisciplinary OHS research and education experience, aiming at providing students an opportunity to learn from each other, from a multi-disciplinary approach of OHS issues, can give them "a sense of appreciation for the forces that shaped and continue to impact occupational health and safety" (Rosen *et al.*, 2011, p. 516). From visiting worksites and working as a team, OHS graduate students were able to have a comprehensive coverage of workplace needs, allowing them to see the hazards faced by the workers, and reflect about how OHS professionals can help to promote healthier and safer experiences at work, acting as transforming agents into the workplaces. At the end, students were able to develop solutions to identified hazards from an interdisciplinary view; for example: to identify and understand the hazards to which workers will be possibly exposed and to provide the suitable and technically feasible conditions to work healthy and safely, occupational physicians, industrial hygienists and safety engineers must work together upon the job-related identified exposure.

Jessup (2007) argues that the transition from multidisciplinary to interdisciplinary teams may help professionals to address workplace changes. Applying her ideas to OHS research, education & practice, one can say that multidisciplinary approaches utilize the skills and experience of individuals from different areas of expertise to complement each other's' skills aiming at reach a common goal, thus providing more knowledge and practical experience than isolated disciplines. Using an interdisciplinary approach, individual discipline approaches are integrated into a single perspective. In such cases, short- and long-term management goals are achieved simultaneously towards the best outcome for the problem. The most obvious advantage over a multidisciplinary approach is its work/worker-centered approach.

Multidisciplinarity, interdisciplinarity, transdisciplinarity, whole-approach and teamwork are words increasingly used and emphasized as current and future needs in research and also in practitioners training and education. In the literature, ambiguous definitions and interchangeable uses can be found, creating sometimes a terminological entanglement (Leathard, 1994). For so, prior to discussion of such needs into OHS universe, some definitions will be presented.

According to Choi & Pak (2006, p. 351), "multidisciplinarity draws on knowledge from different disciplines but stays within their boundaries. Interdisciplinarity analyzes, synthesizes and harmonizes links between disciplines into a coordinated and coherent whole. Transdisciplinarity integrates the natural, social and health sciences in a humanities context, and transcends their traditional boundaries". Additionally, Choi *et al.* (2006) identify that additive; interactive; and holistic are common words used for multidisciplinary, interdisciplinary and transdisciplinary, respectively. Nonetheless, each one of them has a distinct meaning and they should not be interchanged.

A team is usually defined as "a small number of consistent people committed to a relevant shared purpose, with common performance goals, complementary and overlapping skills, and a common approach to their work" (Lorimer & Manion, 1996 p. 15). According to Rosenfield (1992, p. 1351), multidisciplinary teams, "work in parallel or sequentially from disciplinary base to address a common problem". Interdisciplinary teams work "jointly but still from a discipline-specific base to address a common problem" (Rosenfield, 1992, p. 1351). On the other hand, transdisciplinary teams work "using shared conceptual framework, drawing together discipline-specific theories, concepts, and approaches to address a common problem" (Rosenfield, 1992, p. 1351).

To define whether these terms mean the same or different things is not within the scope of this chapter. However, many authors clearly interchange such terms. So, from now on, following Choi's arguments (2006), as the abovementioned terms refer to the engagement of various disciplines working together in some way and to varying extents, the more general term *'multiple disciplinary'* will be used when the degree of such involvement is not clearly specified.

There are many reasons why teamwork consisting of multiple disciplines is desirable in OHS. The same, then, can be said to OHS research and education. Transposing Choi *et al.* (2006; 2007; 2008) point of view into OHS research; education and practice, it is possible to say that life is multiple disciplinary, as well as the workplace hazards. Multiple disciplinary research, education and teamwork must evolve to meet the demands of the new dynamic and constantly changing workplace that cannot be properly met by single disciplines in isolation. In such changing working environments, demands for multiple disciplinarity are emerging as fast as the acceleration of technological development and innovation. Recalling that to address such demands, multiple disciplinary teams, in its simplest version, can be made up of professionals from different areas of expertise, who are able to understand a small part of the real world problems under their own perspectives, but are willing to work together to achieve better solutions for prevention, intervention and promotion of a safety and health working environment (both on research and practice). An effective 'multiple disciplinary' approach can lead to enhanced job performance and work quality and conditions, to improved use of resources and also to improved learning of practitioners and workers.

Occupational safety and health, as a complex matter, requires multiple disciplines knowledge. Consequently, developing solutions to deal with workplace hazards requires a multiple disciplinary approach. The multiple disciplinary (multi-; inter-; trans-) experience and education, therefore, becomes also an important aspect to graduate OHS programs and research. The building of multiple disciplinary relationships at the beginning of OHS professionals' careers, as students, is one of the positive outcomes of a multiple disciplinary learning experience.

Researchers, educators and those who provide OHS to workforce must have in mind that the 'multiple disciplinary' view is a way to remove the dividers between OHS research and practice, thus consisting of an important way to better understand workplace related hazards, injuries and illnesses.

INTEGRATING OHS INTO UNIVERSITY EDUCATION: A STRATEGY?

Some experts argue that to encompass all the necessary aspects of education, training, research and innovation for world of work today's needs, OHS issues should be part of an integrated education strategy.

An interesting contribution regarding this subject is presented in the EU-OSHA report (2010) entitled *Mainstreaming occupational safety and health into university education* and references cited therein. Some aspects related to the inclusion of OSH and risk education as part of university studies will be presented below.

According to the report (EU-OSHA, 2010), more than to ensure that safety and health measures are put in place at workplaces by practitioners, mainstreaming OHS into education has also been a recognized approach to improve overall OHS conditions. Thus, fostering and promoting the safety and health culture, throughout integrating OHS into university level education, became an important area of action. This means to ensure that OHS professionals, directors, managers and supervisors make OHS an intrinsic part of the way that actions are taken at the workplace, more than making sure that workers know how to act safely. It means also to prepare professionals that enter their careers having a clear understanding that OHS is more than an additional task to be fulfilled. Good OHS practices can be easier achieved if such professionals come to the workplace well qualified on a varied OHS related matters, with a culture of precaution and prevention developed during formal education period.

The report argues that a curriculum approach to mainstreaming OHS into education should look for the integration of safety and health into the curriculum not limited to one specific, stand-alone subject. Rather, safety and health should be integrated as a transversal topic, in different subjects, and as a compulsory part of curricula. This means, for example, that all engineering students receive some risk education, not as an optional module or as part of a specialist OHS and engineering degree only. Likewise, all business students should receive some risk education, relevant to their course and as an intrinsic part of it. It is important in order to achieve consistency, sustainability and durability. But making OHS embedded in the curricula is only part of the process. A broader approach should also include, among others, a safe and healthy university environment, for both students and staff, in addition to the learning about OHS as an integral part of any work practice during student life (EU-OSHA, 2010).

Findings of previously published work (EU-OSHA, 2009) concluded that OHS is least likely to be systematically included as an element in courses at university level, and that it is the most challenging educational level regarding the mainstreaming of OHS. The authors pointed out that one reason for this is that universities have high level of autonomy and thus there are no common curricula. They have extensive freedom in what is taught and how it is taught. According to EU-OSHA (2010, p. 16), "the extent and content of OHS education varies greatly in the higher education, and it might not always be proportional to the level of risk that undergraduates could find themselves managing in their future professional working life. Actions to include OHS in relevant courses such as engineering or business studies are therefore generally *ad hoc*, and often dependent on the interest of individual professors or particular advocates within professional bodies".

Higher education authorities must be convinced to embrace the idea of including OHS in undergraduate, graduate and extension course programs. Professors, as well, have to be convinced of the importance and relevance of including OHS as part of their course curriculum. In many occasions, OHS related matters, such as 'risk', are not seen as academic subjects. At other times, there is a lack of expertise to teach or a lack of suitable and relevant teaching resources on OHS for different university level courses (EU-OSHA, 2002a; 2010).

After examining concrete examples on attempts to embed OHS into university, some of the EU-OSHA (2010) report achievements are described below. In general, from the observed practical reality of universities, OHS is more likely to appear as a separate module rather than being truly embedded with course curricula. Only in a few exceptions OHS were embedded in broader strategies. The length of OHS modules offered as part of university courses also varied considerably and OHS courses involving other engineering disciplines are more likely to be optional. OHS modules on business courses are either non-existent or very limited. The lack of awareness of risks in the administrative professions, such as work-related stress, and the notion that OHS is a technical subject may mean that staff and students from administration sciences are less likely to see OHS as an issue for them, even though business and administrative decisions and actions can affect the OHS at the workplace. Technical faculties are more likely to focus on safety issues than occupational health or work organizational issues, which are more often taught by faculties of medicine.

To cite a practical example, the report presents that, in spite of both professors and students usually recognize that practical work in a laboratory is perhaps the most hazardous part of university activities, they still need specific instruction and training in laboratory safety. Training, which in turn, must go beyond a simple presentation of laboratory rules, focusing on hazards identification and risk assessment in a way that students develop a more general awareness and skills related to risk and its prevention. After all, is in the laboratory work where students can see the OHS principles and requirements put into practice in their own work for the first time (EU-OSHA, 2010).

There are many other challenges on attempting to integrate OHS into university level education. This includes, for instance, the lack of suitable OHS educational materials for the university level. The materials need to appear relevant and interesting to both professors and students. Existent available materials are mostly directed for workplace training programs instead for academic educators. Support is needed to adapt existing examples of good practice and interventions from the workplace to the university level, as well as promote an exchange of both ideas and concrete tools (EU-OSHA, 2010).

Another challenge for mainstreaming OHS into universities, pointed out on the EU-OSHA report (2010), is the lack of university level teaching staff with OHS expertise and/or active and participatory education skills, as well as the lack of OHS knowledge among academics, whose specialty is not OHS related areas. Additionally, professors usually do not receive pedagogical training and do not know active learning methods; only on rare occasions, university professors have OHS teaching qualification.

EU-OSHA report presents that introducing practical, active learning methods for OHS in a learning environment dominated by theoretical learning methods, addressing large class sizes, is also a challenge. For a practical topic such as OHS, the development of skills, attitudes and behaviors, as well as theoretical knowledge, cannot be achieved through the use of lectures only. In such case, alternative active learning, as visiting and working in real workplaces, is an important and effective way to improve the safety and health awareness of students. Practical aspects of working tasks and workplaces are difficult to replicate in the classroom, even using resources such as videos to take the student out to the real-life situation. On the other hand, it requires more resources from universities and perhaps a change in mentality by some about how learning at university should take place, as well as a rethinking about the specific skills required on the part of educators (EU-OSHA, 2010).

To the authors of this chapter, one of the most relevant and challenging aspects described on the aforementioned report is the undervaluation of OHS issues among the academic staff. OHS is not seen as priority for learning, but mostly as an obstacle to research, especially when professors are not convinced or not aware of its relevance.

Very often, the university environment itself does not have an OHS culture. The so-called 'thinking OHS' is not an intrinsic part of its daily actions. It is because most universities are not aware of the OHS issues on their campus. In these cases, budgets, time and the availability of resources are always tight, which can push OHS down the list of priorities both for student education and for management and administration. As a consequence, many obstacles may arise to the university safety department to put in place an effective management system for OHS (EU-OSHA, 2010).

The same report highlights that, on the other hand, like any other employer, universities have to run themselves safely and comply with OHS legislation, which means to keep a healthy and safe working for professors, general staff and students at last. Additionally, universities are a very complex environment, comprising many different cultures and languages, employment relations and working patterns, which make the implementation of any OHS management system more and more difficult.

Such scenario should be used as a learning opportunity for professors, students, and technical, administrative and support staff. Working together towards a safe and healthy university environment may be a very effective way to raise awareness among all of them and embed OHS education into students' life in a very practical way. Such approach would combine OHS education with OHS legislation compliance fostering the development of knowledge, skills, safe attitudes and behavior in all of those involved (EU-OSHA, 2010).

Whether embedding OHS into university education could be an additional strategy to cope with the demands posed by the changing world of work, and we believe it is, there is much work to be done. Several steps based on real experiences could be adopted to assist OHS integration into education at universities, such as the cooperative work with individuals and institutions clearly committed to OHS issues, as well as with professional associations related to graduates' curricula. Partnerships with institutions as universities; industries; research institutes; safety authorities and insurance companies could be helpful to the exchange of experiences and to the sharing of learning resources. Cooperating companies could provide, for example, the integration of students with OHS activities – e.g. enabling student visits to workplaces and providing lectures or placements. The cooperative work with business schools is an additional way to include or improve OHS approach at management and administrative related courses, where a lack of training and knowledge about these issues is observed (EU-OSHA, 2010).

Additionally, computer- and internet-based teaching resources could be used as complement of what was taught in the classroom and as a distance learning tool. Academics, in this context, should be provided with assistance to understand and effectively use the teaching materials. To adopt an active-learning strategy could be effective as well at the mainstreaming of OHS into university, by using problems from real cases, presenting methods to solve them; and by giving previous safety instructions before a practical work, in a way that reinforces a prevention culture amongst the students. In this regard, it is also important that the university or those responsible for disciplines/courses are open to discussions about – for example – changes in graduate curriculum. A broader approach to promote and facilitate a safe and healthy environment at the university as a whole could engage, in addition to the students, the academic staff and workers into the OHS issues, as already mentioned (EU-OSHA, 2010).

Another important OHS promotion aspect that should be taken into account is the learning from successful experiences of mainstreaming OHS into academic education (adapted to university) and the use of this knowledge as a practice to train young workers, as well as the encouragement of employees and employers to take OHS issues and knowledge as priorities - including for recruitment purposes (EU-OSHA, 2010).

Some of the described steps could be especially helpful in the cases where there is not enough time or opportunities to present the OHS issues as separate modules, due to the natural limitations on the undergraduate time and required curriculum demands. But it is noteworthy to mention that the 'modular learning' and the development of a specific and directed module for OHS is an important step to the mainstreaming of this subject into the graduate curricula. At least, although it may seem difficult to promote a concern about OHS in environments such as universities, the adoption of the presented steps (considered as 'success factors') could represent a future improvement in work conditions, especially in the context of the changes at workplaces and work conditions (EU-OSHA, 2010).

Mainstreaming OHS into university education is evidently not the only option to raise the concerns; knowledge; practice and skills about OHS (not all the workers have access to university teaching, for example). It is, instead, only one of the strategies to increase the awareness of OHS as an essential issue at workplaces and the adoption of a culture of prevention among everyone in the workplace. It is important to unite the university-level education with other methods, such as the approach of OHS at basic education and the continuous training of practitioners.

BEYOND KNOWLEDGE

The world of work has remarkably changed during recent years, mostly with regard to workload; years of employment; types of labor contracts; work organization; and working conditions (Papadopoulos *et al.*, 2010; EU-OSHA, 2002b; Storrie, 2002). These changes may include "increase of retirement age, increase in daily and weekly working hours, 'deregulation' of working hours, temporary and part-time work, labor leasing, outsourcing, subcontracting, self-employment and down-sizing of enterprises, increased workload and time pressure for workers" (Papadopoulos *et al.*, 2010, p.945).

In order to keep in pace with the global changes in politics; economy and culture, OHS professionals should develop competences for the broad range of challenging tasks they are likely to face.

A person's competence can be described basically by two words: knowledge and skill. It is clear that, more than knowledge, OHS professionals must develop their skills. On one hand, according to the Oxford Dictionary (2014), knowledge refers to "theoretical or practical understanding of a subject". Complementarily, Collins dictionary (2014) defines knowledge as "awareness, consciousness, or familiarity gained by experience or learning" - this could happen, for example, through books, media, encyclopedias, academic institutions and other sources. On the other hand, skill is the "special ability in a task", acquired by training (Collins dictionary, 2014). Therefore, it refers to the practical application of an acquired knowledge within a specific context, to get the expected results. Both knowledge and skill are required to play, in an appropriate way, any professional role (Kumar, 2013).

In previous published articles, attempting to assess outcomes of OHS education, Brosseau & Fredrickson (2009) grouped a core of 29 competencies into six categories: recognition, evaluation, control, communication, behavior and management. All of them were designed to describe the knowledge and skills expected of OHS students and professionals. These professionals need to acquire knowledge and skills according to their field of study and their role into an organization. To illustrate, industrial hygienists need to develop knowledge and skills clustered in each of six categories, but those related to recognition; evaluation and control are clearly most important to succeed. This sub-chapter, however, is not aiming at ranking or associating competencies with professions. Rather, it is aimed at pointing out and briefly describing skills that are common to every OHS professionals in all core disciplines. In the next paragraphs, some of such listed aspects will be brought up into discussion, within this chapter's context.

According to Brosseau & Fredrickson (2009), recognition skills include take into account the influence of cultural and social factors on OHS related matters; associate hazard exposures and possible adverse health effects; recognize and associate hazards with its specific sources and processes; and understand physical, chemical and biological aspects of hazards sources.

Evaluation skills, in turn, encompass to plan and conduct research; gather, manage and analyze data; understand research and use research methods and results; evaluate aspects of exposure assessment, dose response and risk characterization – including the planning and implementation of exposure assessment strategies; have a basic understanding about sampling and its relation with the evaluation of exposure and controls; prioritize hazards and exposures, also focusing on what is necessary to its elimination or control (Brosseau *et al.*, 2009).

Control skills are mainly related to the design and implementation of health promotion, work interventions and work environment modifications; recommend, evaluate and implement controls; and select the most suitable control methods for specific exposures (Brosseau *et al.*, 2009).

Every OHS professional must have an understanding of an organization's power struggles and politics, as well as its cultural and social values. It is a recognition skill that makes it possible to work in accordance with the political environment, and without compromising the research or any other type of intervention. Such understanding allows the professional (researcher or practitioner) blend into the surroundings and be virtually invisible. It enables, for example, the research setting to have minimal disruption. Being aware of organization's characteristics may help the professional to make recommendations, interventions, or even select the most appropriate control methods for a given workplace (Teusner, 2010).

Design and initiate research, an evaluation skill, may be related with the adoption of new research methodologies to improve results and performance, as for example the insider research. Despite not so outstanding, such methodology can be a very useful approach to be applied to OHS matters, due to practitioners choose to research their workplaces to make improvements to OHS systems and practices. Such methodology maximizes insiders' knowledge, allowing better understanding the application of the scientific findings in a more effective way, and, consequently, enabling the design and implementation of changes in the work environment (Teusner, 2010).

Communication skills, according to Brosseau *et al.* (2009), involve, among others, to communicate effectively with team members and with other OHS health professionals; to write well and to design and deliver education at all levels. These skills must be developed by all of those working on OHS subjects. It is the ability to make OHS goals and actions clear to workers, looking for transparency and openness in their actions. Communication still refers to the ability of influence, negotiate and convince varied audiences. Without effective communication skills, OHS practitioners cannot, for instance, properly gather information, transmit recommendations, convince people to make changes, and even translate scientific data to a form that is suitable in an organizational context within workplace. Such ability enhances professionals' credibility among their workmates (Teusner, 2010).

Training and development of workers can be interpreted as education design and delivery, another important skill to be developed. It is one of the most important aspects in the management of people at work, once it improves skills and quality of job performance. Continuous changes are better accepted by those workers who are trained under a continuous process. OHS practitioners or any other involved in workers training and education must make efforts to bring also workers skills, knowledge and abilities up-to-date. It is an opportunity to make them fit into new changes taking place both within and outside their work organizations (Okafor, 2007).

Occupational health & safety practitioners must be able to understand the workforce's diversity in order to develop and deliver key training programs, as well as to propose and implement interventions to address the wide range of beliefs, customs, or practices that negatively impact health and safety at work (NRC, 2000). In other words, training and education must be adapted to the workforce characteristics. The wide diversity of cultures, mental sets and corresponding behavior of the workforce at a large must be taken into account.

Design and delivery of education allow OHS practitioners to choose and prioritize specific training courses for workers. Such development plan must be drawn, being aware that most developmental activities carried out on-the-job are intended to build defined knowledge and skills that foster growth in required competencies.

The aging of the workforce, that means a great many people working past the traditional age of retirement, can be used as a good example for that. Research findings show that, despite their age, older workers present several desirable characteristics, such as lower turnover; lower absenteeism; and lower accident rates. This professional profile, combined with an up-to-date skill, constitute a fundamental resource to employers. This means that training older learners is becoming increasingly important for organizational effectiveness, once such workers are demanded for new knowledge, skills, and abilities to perform most jobs, as technologies advances and rapidly changes (Callahan, Kiker & Cross, 2003).

Nevertheless, training older learners involves considering the influence of several physical and cognitive characteristics of the aging process, such as the lowering of cognitive response times and also reduced concentration and memory, which in turn, may result in reduced registration of training information. Those responsible for training such workers must know these facts and be specifically prepared to deal with them (Callahan *et al.*, 2003).

When designing training programs for older learners, motivation; structure; familiarity; organization; and time must be taken into account. According to Callahan *et al.* (2003, p.666), "research on training older learners suggests that they need to know why they must learn a task prior to the training experience or their motivation to learn may be compromised". Due to their accumulated experience and know-how, adult learners are essentially focused on master a given task and solve specific problems. This is the reason why they need to be informed and convinced before undertaking learning. So, learning effectiveness can be particularly improved through their active participation on task performance. Instead a passive approach, training programs must be based on problem-centered issues and more focused approaches. In such context, information must be introduced in a logical order, following a difficulty level sequence. Training program must be built on a current knowledge base and instructions related with memory-building should precede content information (Callahan *et al.*, 2003).

As older learners take longer to fully absorb the content of training materials, course environment must provide to learners the conditions to overcome all training activities. Regarding time, self-pacing is an interesting way, probably the best option, to harmonize the individual differences in learning, which naturally occur with aging. According to Callahan *et al.* (2003, p. 668), "self-pacing allows older learners to internalize the importance of the training and provides them ample time to complete the task and master the training content", which certainly improves the effectiveness of the training program.

Following Brosseau *et al.* (2009) competencies categories, management skills, according to the author, involve effective leadership and teamwork. It means that the professional (researcher or practitioner) must understand the social and cultural context of a given workplace (in other words, the organization's custom). They must have in mind that an organization is made up of a collective of people who share similar values or basic assumptions that have been learnt, producing a culture of a shared or agreed way of doing things (Teusner, 2010). The intimate knowledge of the context of the organization, the way such organization understands the formal and informal approaches taken and, besides, an intimate knowledge of their own culture are also extremely important factors on leadership. Leadership involves, among other characteristics, personal commitment for coping, managing and struggling with situations. Team leader must have good ideas and perspicacity, acknowledge the importance and requirement for multiple kinds of expertise, know how to integrate information and work to maintain everyone on the team. Team leader must also know how and when integrate professionals, despite the fact that, although multi- or interdisciplinary teamwork is desirable, such integration is not always feasible or necessary (Choi & Pak, 2007). Previous published study shows that issues such as management and leadership become crucial for safety performance in this changing working scenario (Mearns & Yule, 2009; Herrera, 2012).

Behavioral skills, according to Brosseau *et al.* (2009), include awareness of diversity in social and cultural beliefs; effective work on an interdisciplinary team and stay current in one's field of practice. It also encompasses to take initiative and to be creative and collaborative to develop solutions and to make an assessment about what is practical and realistic to study, considering resources and time required (Teusner, 2010). To work effectively in a team, OHS professionals must have maturity and flexibility, personal commitment, a common goal and shared vision (Choi *et al.*, 2007).

All of the mentioned changes in the world of work also increase the importance of a behavioral skill called as resilience. *Resilience* has become a popular *buzz*word in OHS area. For some, it seems to represent "an answer to the threats and uncertainties associated with the fast paced changes of modern society" (Hovden, Albrechtsen & Herrera, 2010, p. 954). It refers to the ability to manage the unforeseen; how a professional or a team becomes prepared to cope with unexpected outcomes. The word resilience has its origins in the Latin word *resilire*, which means to leap back or to rebound, and refers to a person's ability to recover from challenges or disrupting events (Woods, 2006). But, more than a person's adaptive capacity to confront and bouncing back from disruptions; constant changes; and pressures, resilience also means the ability to stretch, to move forward and emerge stronger from such situations.

The resilient OHS practitioner looks for opportunities in problems, has always a positive attitude, seeks to overcome difficulties and learn from mistakes. Thus, being proactive and adaptive, such professional can be able to respond effectively to several changes in today's complex working environment, ensuring the delivery of an up-to-date OHS service to the workplace and to the workforce. Such professional can contribute to build a resilient workplace and, at last, a resilient workforce. This workforce, in turn, can be described as "a group of employees who not only are dedicated to the idea of continuous learning, but also stand ready to reinvent themselves to keep pace with change; who take responsibility for their own career management; and, last but not least, who are committed to the company's success" (Waterman, Waterman & Collard, 1994, p. 88). Accordingly, the resilient workforce also implies in commitment to OHS systems and policies.

A resilient workforce is aware of their own strengths and weaknesses; understands the skills and behaviors needed for the future; has the readiness and ability to adapt to new workplace requirements in a flexible and fast manner. In such context, employees become committed to the OHS strategy and shift the eyes from navels to the collective safety and health. At last, only a resilient workforce can succeed in a scenario where the required skills are rapidly changing (Waterman *et al.*, 1994).

Resilience is not something that a person is either born with or not. It is improved and can be learned as people acquire self-management skills and more knowledge, across an entire life span. Feeling in control; having a positive view of yourself and having confidence in your strengths and abilities are factors that contribute to resilience development (Gilbert, Sweet, Gazit & Youngelson, 2010). On the other hand, resilience increases the importance of self-reliance to both OHS professionals and workforce. Self-reliance, in turn, means the ability to rely on one's knowledge, skills and motivation. It requires "a continuous openness of individuals towards new learning and self-adaptation (life-long learning)" (Wilpert, 2009, p. 730). This means, among others, stay current in your own field of knowledge and practices, keeping up with new knowledge and with new innovative ideas. It will reflect a self-reliant teamwork and, at the end, a self-reliant workforce. This means that, more than life-long learning, OHS professionals are required to learn how to learn during an entire life span. More than technical expertise, there is an increase of the importance of more general qualifications, such as those described above. It is clear, then, that new questions emerged and remain unanswered: who prepares for such qualities? How can this lack be overcome?

BRIDGING KNOWLEDGE AND PRACTICE

In the last decades, society has watched a fast technological development that led to the introduction of new working practices, and consequently the introduction of new health and safety hazards intrinsic to both process and products. And it is *common sense* that the changes previously reported impacts and will continue to deeply impact the workplaces, lowering OHS conditions.

Up to now, this chapter has briefly approached the major changes observed in the world of work, education and training needs, professionals skills requirements, and also a possible strategy to enhance the '*thinking OHS*' attitudes and behavior. Various national and international resources were used to expose these topics: scientific papers, international agency's reports and books. Some of them were used as general background resources, explicitly referenced throughout the text, such as EU-OSHA reports (1998; 2000; 2002a; 2002b; 2002c; 2002d; 2009; 2010) and the USA National Research Council book (NRC, 2010).

Research findings usually point towards the great need for an up to date professional. As previously mentioned in this chapter, a rethinking is needed, once the changes in the world of work have affected OHS professions structures as it was formerly known. OHS professionals' competencies are at stake.

Several modifications must be introduced in curricula at different levels of education, which now must be adapted to meet new workplace demands. The education approach must be reviewed; a transdisciplinary approach must be adopted when possible, keeping in mind that it does not mean a '*moving away*' from single-, multi- or inter-disciplinary approaches. Mainstreaming OHS transversally into university education, more than a teaching/learning strategy, is a way to raise awareness among professors and students, as well as technical; administrative and support staff. It is an opportunity to students to learn from practice and perhaps it is an opportunity to enhance the prevention and safety culture expressed not only in terms of knowledge, but also as skills, fostering the development of knowledge, skills, safe attitudes and behavior in those involved. It is a way to enhance OHS culture in an environment where such concepts are far from reality and under evaluated. To accomplish all the OHS contents that must be approached, alternative learning methods must be developed and implemented.

Continuing education to promote lifelong learning is one of the tools to bridge academic knowledge and practice, thus improving the understanding and skills of OHS professionals. Investments on research through Ph.D. and masters programs must be maintained and extended; students must be funded. The evolving demands to meet the challenges of changing workplaces require a much broader perspective than that taken until now.

Professionals must work together to achieve high quality OHS research outputs, in addition to the delivery and the implementation of services. OHS professionals must be prepared to disseminate good practices of work and also a preventive and precautionary perspective to deal with new emerging issues.

The research priorities must be driven by the introduction of new technologies that comes along with new and emerging risks, in addition to traditional risks. Demographic changes and new forms of work organization must also be taken into account. The main goals are to strengthen OHS programs at workplace, as well as to focus OHS strategies on a global and integrated promotion of a healthy work.

Partnerships and research collaborations are crucial to share knowledge and resources focusing on the improvement of working conditions and quality of work. For so, a proactive dialogue is needed to propose adequate solutions and interventions, which in turn will help to achieve health and safety goals within the workplace. In sum, a critical mass of OHS researchers, educators and practitioners must be tailored to best cope with the challenges posed by the changes in the world of work.

Nevertheless, it is important to highlight that former and current research approaches usually do not fit to the intrinsic complexity and dynamism of today's working life. From now on, OHS researchers should focus on practical research, prioritizing intervention studies. Fundamental research must aim at solving important applied problems.

As authors, we believe that bridging the knowledge gap from research and education to practice is the huge barrier to overcome. It is the biggest challenge to be faced by researchers, professors, students and practitioners. The immense gap between research and practice is debated by several authors (Salazar, 2002; Kerner, 2008; Ferguson, 2005; Glasgow, Lichtenstein & Marcus, 2003). The reasons for the failure of translating research knowledge into practice are many, including political, social, cultural, scientific and economic, among others.

As stated by Salazar (2002, p.520), "in most organizations, standard operating procedures and behavioral norms are the major influences on workplace practices; scientific evidence plays a minor role". Salazar (2002) points out many bottlenecks to application of research outcomes. Firstly, practitioners may not always realize the importance or practical application of research studies in the workplace. Practitioners themselves must be skilled and requalified for that. Secondly, very often, the implementation of new practices requires more oversight than practitioners could provide. Thirdly, practical application of research results demands a close collaboration of several individuals in a committed teamwork, which is not easy to achieve in the current dynamic working environment, mainly characterized by extensive turnover.

Kerner (2008) asserts that the most important challenges to bridge the gap between research and practice are to strengthen partnerships; to adapt and harmonize terminology, avoiding misuses; and to define the meaning of the word 'evidence' across the research, practice, and policy areas.

Ferguson (2005) points out that due to globalized political and economic scenario, organizations, struggling to survive, are mostly focused on daily routine activities and operations. Efforts to betterment of OHS conditions, based on innovative research results, demand investments on human resources and time to reflect and rethink future directions. That is, shift thoughts from 'thinking on now' to 'plan the future'. In addition to limitations in time; financial resources and managing support, the author argues that there still are three cultural factors which can hinder the integration between research and practice: institutional differences, such as when companies have different goals, intended beneficiaries and ways of working; communicative differences, due to the variety of vocabulary among audience, according, for example, to the area of expertise and level of education; and philosophic differences, mainly caused by different interpretations to the knowledge definition (varied *epistemological* views).

Another interesting point of view about the reasons for the practice-research gap is presented by Glasgow *et al.* (2003, p. 1261), which approaches that this gap is formed due to "several interacting factors, including limited time and resources of practitioners, insufficient training, lack of feedback and incentives for use of evidence-based practices, and inadequate infrastructure and systems organization to support translation". The inference that research's effectiveness is a direct consequence of research's efficacy contributes to such gap. Efficacy research is conducted under standardized conditions. Variables such as context, environmental conditions and studied population are considered under a narrower perspective. Effectiveness research, in its turn, is conducted under 'real-world' conditions, aiming at to understand variations in results across heterogeneous conditions and populations, producing more robust and long-term results (Glasgow *et al.*, 2003).

With all these obstacles to surpass the gap between research and practice, it is essential to develop strategies to cope with the diversity in the world of work and to encourage the management support and the provision of resources for promoting an effective understanding between researchers and OHS practitioners. Without the combined work of these professionals, it is not possible to create an effective approach to apply the knowledge in real workplaces and work conditions. Consequently, without this it is also not possible to reach and engage workers in OHS issues and the research findings will poorly be implemented in reality - since they will fall outside the context of practical work.

Don Norman, Professor at Northwestern University, wrote about bridging gap in his bimonthly column for *Interactions*. In his paper *The Research-Practice Gap*, Norman (2010) states that, without a word of criticism, the knowledge and skill sets required of researchers and practitioners groups differ considerably. Transforming a research into practice is an intricate challenge, for which neither researchers themselves nor OHS practitioners alone are skilled. Skills for conduct a creative research are completely different from those required to put interventions in place. Few are those who have the breadth and depth of skills and knowledge to practice and, concomitantly, explore research results of several OHS related disciplines, making use of them. This is the reason why the research-practice gap is a worldwide phenomenon so difficult to overcome (Norman, 2010).

For scientists, hypotheses, evidence and conclusions are clear linked and related to each other. For most practitioners of any profession, daily work is based upon the best practices known. While scientific research is conducted under a controlled environment, the real workplace conditions are complex and sometimes uncontrollable (from the scientific point of view), contradicting the logical and well-established assumptions of scientists and hindering the so-called scientific rigor. Norman attributes a further enhance of the research-practice gap to the researchers who often conduct their studies without knowing if they will have an effective application in real work environments and conditions. The author argues that even if the research is aimed at finding a solution to a practical problem, there is still a difference between the knowledge and the skills required by the ones who conduct the research and the ones who apply it at workplaces - which effectively seek to translate the research results into a practical scenario, taking into account reliability; feasibility; and affordability in real workplaces. On the other hand, as also pointed out by Norman, practitioners are not trained with focus on scientific research.

According to Norman (2010), Donald Stokes's book, *Pasteur's Quadrant*, suggest a solution to this dichotomy between research and practice, arguing that research is more effective when carried out into a context, or in other words, when the research is performed aiming at specifically solving problems in a real situation. Stokes (1997) argues that more substantial outcomes lie on the search of elemental knowledge within a specific context. This is the case, for example, of the smallpox vaccine development. From a real situation, Pasteur developed a fundamental scientific research and applied the scientific knowledge advances back to the real problem (Norman 2010; Stokes 1997). A way of bridging the research and practice gap is presented by Norman, which stated that a third discipline must be inserted between research and practice: a discipline of translational development. This intermediate field was considered by the author as needed in all arenas of research. Translational developers would act in between research and practice, translating scientific research outcomes into reliable and useful practical results. Similarly, the author defends that translational developers would clearly translate practical problems and concerns, supporting researchers to come up with need-based solutions to practical issues. In fact, he argues the need for translation in both directions: from research to practice and from practice to research.

Moving towards this direction, the National Institute for Occupational Safety and Health (NIOSH) has adopted the term *Research to Practice* to describe applied research focused on practical solutions to reduce occupational injuries and illnesses. It is reported as an initiative focused on "the transfer and translation of knowledge, interventions and technologies into highly effective prevention practices and products which are adopted into the workplace" (National Institute for Occupational Safety and Health [NIOSH], 2011). It is also described as an alternative way of developing research and delivering results to the interested parties, who by making use of such outcomes could reduce occupational injuries; illness and fatalities into the workplace.

The transfer of research into practice and the sharing of research outcomes with OHS practitioners and with those who are in charge of work/workplace organization and changes become now a crucial area to deal with working society needs regarding OHS issues. The bridge between practice and research must be supported by user-friendly suitable material, yet to be developed. It requires long-term commitment among researchers, OHS practitioners, policy makers and managers.

To shrink the research-practice gap, several challenging demands should be addressed. In this regard, Sorensen & Barbeau (2004; 2006) define some research priorities, including the identification of strategies to promote the dissemination and also the sustainability of evidence-based programs. Both researchers and practitioners must be sure about the effectiveness of any evidence-based intervention programs, before its implementation. Only then, they will be prepared to make employers and workers willing to accept, accomplish, disseminate and perpetuate interventions. Equally important are studies on efficacy and effectiveness to assess the outcomes of interventions; health behavior changes; and intervention implementation cost assessment, among others.

Rosenheck (2001) argues that organizational process related matters are still understudied, and potentially a barrier to link research and practice. Among others, the author believes that efforts should be made to link new working approaches to existing policies and customs; that decision making process should involve many different stakeholders; and that the perpetuation of newly introduced practices can be achieved by the creation of 'communities of practice' into the organization.

Kerner (2008) argues that complex issues regarding the meaning of scientific research for non-scientist audience, as well as methods for disseminating and implementing research outcomes, should be addressed to properly merge scientific findings into practice.

Colditz, Emmons, Vishwanath & Kerner (2008) argue that to overcome the research-practice gap, multilevel approaches to research; dissemination and metrics are required to inform academic appointment and promotions. The abovementioned authors state that "moving beyond funding that stops and starts with grant cycles is a key issue from the community perspective to ensure continuity and improved health. Transdisciplinary approaches that cut across disciplinary boundaries to develop shared conceptual frameworks may help speed the integration of research with practice" (Colditz *et al.*, 2008, p. 144). The development and support to disciplinary teams, through identifying and implementing structural changes, are encouraged. The authors additionally ague that changes in the way which scientific advances are structured and implemented into practice will help in the improvement of the safety and health conditions.

Ferguson (2005), on the other hand, presents some building blocks to help surpassing the research-practice gap: to know each other, acknowledging the diversity and differences at all levels; respect and patience; recognition that scientific knowledge depends on practical knowledge (and *vice-versa*); fostering of a mutual frame of reference; building of incrementally partnerships between research and practice, with equal commitment among the parties; assurance that there is a broad institutional buy-in and allowance for mistakes, learning with them.

Glasgow *et al*. (2003) suggest four specific changes to bridge the gap between effectiveness and efficacy researches and approximate theory to practice: increased attention on moderating factors, which narrow the research robustness (in both efficacy and effectiveness researches); realization that efficacy studies are not enough by themselves to actually impact on public health; inclusion of specific criteria in authors guidelines from scientific journals, as a way to require researches with proper external validity and increase the financial support for studies on moderating variables; external validity and robustness.

Until now researchers have been almost solely responsible for trustworthiness of their findings. From now on, researchers have also to be aware of the usefulness of their findings, in an initiative that attempts to overcome the barrier of knowledge practical application, keeping in mind that, to be useful, research results should be clearly reported, comprising relevant issues to practitioners. Research should address practitioners' questions and practice challenges, also involving these professionals in the research.

Focusing on comprehensive interventions resulting of collaboration with practitioners could represent the engagement of them with the research. As argued by Baker, Israel, & Schurman (1996, p. 175), "occupational safety and health and worksite health promotion practitioners need to develop more comprehensive interventions and rigorously evaluate these programs to determine if they are more effective than programs with a more narrow focus" This, together with the help from the translational developers (bridging professionals), should result in a joint responsibility between researchers and practitioners, and now take a role to improve OHS conditions at workplaces.

Additionally, the integration between research and practice will only succeed if there is commitment of those who plan; implement and evaluate OHS within the work activities. The management is also fundamental to surpass this gap, it should support and provide the resources to conduct and apply OHS researches, knowing that the results will come mostly at long- and medium-term, rather than instantly. In this respect, it is also important to highlight that OHS research should not be approached in a broader way. The knowledge produced by research needs to take into account the specificity of the work environment studied, in order to successfully apply the studies in 'real-world'.

To summarize, notwithstanding the availability of resources to develop the knowledge and the scientific research in OHS area (as, for example, creation; adaptation and update of undergraduate and graduate programs; research funding support and a diversity in number of trained OHS practitioners), it is not possible to reach the end of the chain (that means, the *real world* workers) if the knowledge obtained from research and training is not applied to the practical reality of the changing workplaces and work relations. It is necessary to surpass the gap to maintain adequate safety and health conditions at work, avoiding occupational illness and injuries. Without the bridge between research and practice, which results in the gathering between researchers and workers as a whole, the adoption of a prevention culture, the awareness of good practice in OHS and the health promotion become difficult to apply in real workplaces.

CONCLUSIONS

Along with the industrial and globalized market development in the last century, many issues have emerged as new frontiers in Occupational Health & Safety. Such frontiers can also be understood as new challenges for OHS researchers and practitioners.

Occupational Health & Safety, as it has been conducted, was based on the paradigm of a person working during an entire working life for a single employer, at the same workplace, performing a simple and well defined task. Nonetheless, changes in the employment relationship, in the workforce demographics and in the work organization mean that a growing number of workers no longer fits this traditional paradigm.

Dealing effectively with aging of the workforce, for instance, means dealing now with the differences in both physical and cognitive workers' characteristics, as well as with the interaction of disabilities and chronic diseases with workplace demands. It is known, for instance, that older workers take longer to adapt to new and demanding workplaces and processes. Labor migration across national borders, in turn, requires OHS professionals to develop novel communication skills to reach and integrate multicultural groups.

Continuous alteration in technologies, tasks, working hours, working shifts, working patterns and work load, make it extremely difficult to plan, implement, control and monitor health and safety measures, programs and interventional strategies. For instance, in a changing working environment, formerly adopted exposure limits might be outdated. The traditional view of OHS interventional, preventive and surveillance approaches is no longer useful in most situations. Nevertheless, the OHS professional should be able to propose innovative and feasible approaches, which mean that, in spite of being innovative, cannot be detached from organizations' reality.

Faster industrialization, adoption of information technology and automation have changed the work processes and have provided businesses the capacity to initiate, eliminate, change, or transfer several activities, leading to job redefinition and also work redesigning. This has also led to a rise in the number of workers exposed to traditional and new occupational risk factors. As a consequence of fast technological development, OHS professionals are facing emerging hazards, emerging risks, risks whose characteristics are not completely understood, or are even intangible/intractable. Effectively protect workers who are exposed to emerging hazards is not an easy task, when risks associated with such hazards are nothing but suggestive.

Changes, briefly summarized above, have been witnessed worldwide. They have profoundly affected not only the working life, but also the structure of OHS professions, and the role of OHS professionals within company's, workplace, workforce, and at last, society.

Despite improvements in working processes and products, OHS professionals are and will be a necessary part of doing business worldwide, as researchers; educators; or practitioners, in both private and public sectors. The economic, political and social changes keep affecting how and where they work; even so, there is no lessening of their need.

Nevertheless, the competencies these professionals must acquire and develop to keep in pace with the rapid changes in the world of work are at stake. It has been challenging to rethink training and education for creating a competent critical mass of OHS experts, able to understand and to deal the society's needs in the changing world of work. The current OHS professional must be aware of OHS subjects' relevance, as a more integrated approach towards a broader goal of wellbeing of workers during their entire working life. They must also be prepared to present OHS related matters as an essential piece for responsible and sustainable development – showing, for instance that the rapid growing of new technologies can potentially expose an increasing number of workers to emerging and unknown risks.

International agencies and organizations, such as NIOSH and EU-OSHA, have been discussing the changes in the world of work; the challenges to; and also the changes on education of OHS professionals, as well as research priorities areas. In this regard, the USA National Research Council made a comprehensive analysis towards the needs and ways to be followed in order to cover all the new aspects that must be approached, improved and inserted into OHS core professional curricula, as well as other closely related OHS professions.

Redesigning research and learning methodologies – from *multi-* to *inter-* or *transdisciplinary* – is a very important issue, due to the need of integrated and more holistic solutions to be implemented on new dynamic workplaces. Multiple disciplinary approaches are clearly required for develop suitable and effective solutions to the complex and multifaceted problems of today's working environment; therefore, whether *multi-, inter-* or *transdisciplinary* will be defined according to the matter in concern.

Mainstreaming OHS into university education, as extensively discussed on the EU-OSHA report, is a smart strategy. It is vital to meet today's needs, to improve understanding; attitudes and behavior, contributing to the strengthening of an occupational prevention culture from the beginning of the training and education of future professionals.

Changing values at university, allowing students to learn from practice, is a key to change mentality, and perhaps, to instill the preventive way of thinking in such multifaceted environment, where OHS issues are usually under evaluated by professors; researchers; and administrative staff. More than that, it is an opportunity to enhance the value of OHS for more students not directly studying OHS subjects, and, in the end, for the society as a whole.

The educational environment should nurture OHS professionals who must be not only qualified but stimulated to go beyond the established frontiers of OHS, to find new; better and broader ways of building safety and health programs, with a more holistic view than those traditional safety and health paradigm, epitomizing the health-giving properties of doing; being and becoming according to different abilities. The 21st century professionals must have in mind that ensuring protection of workers' health and safety means more than the retention of work ability. For so, appropriate methodological approaches are required to address emerging OHS issues, which includes long lasting research and life-long learning.

Nevertheless, the utmost question that emerges from the discussion presented above is the quest for the balance between *breadth of knowledge* and *depth of knowledge*. Currently, there is a tendency to believe that education tailor OHS professionals to have both broad and deep knowledge of OHS matters. It is the belief on training OHS professionals to be generalist and also a specialist (sometimes multi-specialist). The generalist can see the big picture. Theoretically, such professional is able to perceive a holistic view rather than a narrow perspective. It is important for recognizing problems, but not to identify solutions. On the other hand, the specialist is the most capable one to solve complex and specific problems. It cannot be expected from all specialists to see the broader picture. In a dynamic changing society of nowadays, however, to really succeed, OHS practitioners must reach breadth of knowledge and also, the depth in some area, which means being a generalist with some skills and knowledge deeply developed. OHS researcher, on the other hand, must initially have specialized competencies, and to effectively develop applied and focused researches. In the real world, they must make efforts to see the big picture.

Formal education, continuing education, curricula adaptation, generation of knowledge, learning methods, all of this must be reviewed, adapted and implemented. It is the starting point to face the challenges posed by the mentioned changes. However, currently and from now on, something more than mere technical expertise must be developed by OHS professionals to deal with diversity, complexity and rapid changes previously discussed: skills. It becomes crucial to OHS professionals.

As previously mentioned, one's competence can be described by knowledge and skill. It refers to the practical application of an acquired knowledge within a specific context, to get the expected results. In other words, without developing skills, the acquired technical expertise will not be effectively applied, whether as researchers or as practitioners.

Among the skills approached, resilience is certainly one of the most important. The adaptive capacity to confront disruptions and constant changes in the workplace and workforce is imperative for OHS professional to respond effectively to the dynamic and complex working environment. To be resilient, nevertheless, the professional must be self-reliant, which in turn, requires a continuous openness towards life-long learning. In other words, these professionals must stay abreast of current information needed to stay in the forefront of important workplace issues. It means, then, the introduction of an additional variable: the OHS professional must know how to learn during an entire lifespan.

Regarding this, new challenges arise: who prepares OHS professionals for such? How can the current lack of such skills be overcome? This is a point to be discussed by everyone involved in training and education of OHS professionals. Until now, there is much talk about skills to be developed; very little is said about how to develop or to transmit them. To us, as authors, it is not clear the direction to be followed.

Throughout the twentieth century, many efforts were made in order to improve safety and health conditions in the workplace. Massive studies were conducted on several aspects of OHS, such as work organization, work relationship and risk management models. There was a clear increase of awareness among OHS researchers, practitioners and workers. Nevertheless, despite all the theoretical knowledge gathered, workers still die as a consequence of occupational injuries and accidents, changes continue to impact workplaces, and OHS is becoming an increasingly complex issue everyday.

It is quite obvious that there is a large gap between produced knowledge and putting it into practice. Attempts should be made to strengthen the link between knowledge and practice. After all, the dissemination of research findings and also good practice examples are a way to prevent negative consequences and reinforce positive effects. Being a *translational developer*, adopting Norman's terminology, is a skill to be developed by both researchers and practitioners. Despite being few, in numbers, the research findings point towards the direction of joint efforts, joint commitment and joint research design. It is the biggest challenge to be faced by researchers, professors, students and practitioners. Challenge of which, can determine the success or the failure on the enhancement of OHS conditions.

At the sharp end, little has changed. In few words, the need to protect workers and enhance OHS conditions in all sorts of workplaces remains. Nevertheless, a cultural revolution is needed to keep OHS profession in the forefront, to surpass the frontiers of knowledge. It is now time to rethink the way OHS is approached by governments, universities, employers, OHS professionals and workers. Upon moving into the 21st century, they must undertake a joint commitment to address continuing changes in the world of work, aiming at ensure, above all, workers safety & health in a healthy and safe workplace.

ACKNOWLEDGEMENTS

None declared.

CONFLICT OF INTEREST

The authors confirm that this chapter content has no conflict of interest.

REFERENCES

American College of Occupational and Environmental Medicine. (2014). *Occupational and environmental medicine physicians:* keeping America's workforce healthy. Retrieved from http://www.acoem.org/healthyworkforce.aspx

Baker, E., Israel, B. A., & Schurman, S. (1996, May). The integrated model: implications for worksite health promotion and occupational health and safety practice. *Health Education Quarterly, 23*(2), 175-190.

Benavides, F. G., & Benach, J. (1999). *Precarious employment and health related outcomes in the European Union.* Luxembourg: Office for Official Publications of the European Communities. Retrieved from http://www.eurofound.europa.eu/publications/htmlfiles/ ef9914.htm

Boedeker, W., & Klindworth, H. (2007). *Hearts and minds at work in Europe: A European work-related public health report on cardiovascular diseases and mental ill health.* Essen: BKK Bundesverband.

Brosseau, L., & Fredrickson, A. (2009, May). Assessing outcomes of industrial hygiene graduate education. *Journal of Occupational and Environmental Hygiene, 6*(5), 257-266.

Callahan, J. S., Kiker, D. S., & Cross, T. (2003, October). Does method matter?: a meta-analysis of the effects of training method on older learner training performance. *Journal of Management, 29,* 663-680.

Choi, B. C., & Pak, A. W. (2006, December). Multidisciplinarity, interdisciplinarity and transdisciplinarity in health research, services, education and policy: 1. definitions, objectives, and evidence of effectiveness. *Clinical & Investigative Medicine, 29*(6),351-364.

Choi, B. C., & Pak, A. W. (2007). Multidisciplinarity, interdisciplinarity and transdisciplinarity in health research, services, education and policy: 2. promotors, barriers, and strategies of enhancement. *Clinical & Investigative Medicine 30*(6), E224-E232.

Choi, B. C., & Pak, A. W. (2008). Multidisciplinarity, interdisciplinarity and transdisciplinarity in health research, services, education and policy: 3. discipline, inter-discipline distance, and selection of discipline. *Clinical & Investigative Medicine, 31*(1), E41-E48.

Colditz, G. A., Emmons, K. M., Vishwanath, K., & Kerner, J. F. (2008, March/April). Translating science to practice: community and academic perspectives. *Journal of Public Health Management and Practice, 14*(2), 144-149.

Collins dictionary. (2014). London: HarperCollins. Retrieved from http://www.collinsdictionary. com/

Commission of the European Communities. (2001). *Making a European area of lifelong learning a reality.* Brussels: Author. Retrieved from http://eur-lex.europa.eu/LexUriServ/ LexUriServ.do?uri=COM:2001:0678:FIN:EN:PDF

European Agency for Safety and Health at Work. (1998). *Priorities and future strategies with regard to occupational safety and health.* Bilbao: Author.

European Agency for Safety and Health at Work. (2000). Summary of the changing world of work conference conclusions. *Magazine of the European Agency for Safety and Health at Work, 2*(7). Retrieved from http://agency.osha.eu.int/publications/magazine/#2

European Agency for Safety and Health at Work. (2002a). *Learning about occupational safety and health.* Luxembourg: Office for Official Publications of the European Communities.

European Agency for Safety and Health at Work. (2002b). *New trends in accident prevention due to the changing world of work.* Luxembourg: Office for Official Publications of the European Communities.

European Agency for Safety and Health at Work. (2002c). *Research on changing world of work.* Luxembourg: Office for Official Publications of the European Communities.

European Agency for Safety and Health at Work. (2002d). *The changing world of work: Trends and implications for occupational safety and health in the European Union* [Electronic forum]. Belgium: Author.

European Agency for Safety and Health at Work. (2009). *OSH in the school curriculum: Requirements and activities in the EU Member State.* Luxembourg: Office for Official Publications of the European Communities.

European Agency for Safety and Health at Work. (2010). *Mainstreaming occupational safety and health into university education.* Luxembourg: Office for Official Publications of the European Communities.

Ferguson, J. E. (2005). Bridging the gap between research and practice. *Knowledge Management for Development Journal, 1*(3), 46-54.

Gilbert, D., Sweet, J., Gazit, C., & Youngelson, J. (Writers), & Gazit, C., & Youngelson, J. (Directors). (2010). *This Emotional Life* [Television series]. Arlington: PBS Distribution.

Glasgow, R. E., Lichtenstein, E., & Marcus, A. C. (2003, August). Why don't we see more translation of health promotion research to practice?: rethinking the efficacy-to-effectiveness transition. *American Journal of Public Health, 93*(8), 1261-1267.

Herrera, I. A. (2012). *Proactive safety performance indicators. Resilience engineering perspective on safety management (Doctoral thesis).* Norwegian University of Science and Technology, Trondheim.

Hovden, J, Albrechtsen, E., & Herrera, I. A. (2010, October). Is there a need for new theories, models and approaches to occupational accident prevention? *Safety Science, 48*(8). 950-956.

Jessup, R. L. (2007, August). Interdisciplinary versus multidisciplinary care teams: do we understand the difference? *Australian Health Review, 31*(3), 330-331.

Kerner, J. F. (2008, March/April). Integrating research, practice, and policy: what we see depends on where we stand. *Journal of Public Health Management and Practice, 14*(2), 193-198.

Kumar, M. (2013). *Difference between knowledge and skill.* Retrieved from http://www. differencebetween.net/language/difference-between-knowledge-and-skill/

Leathard, A. (Ed.). (1994). *Going inter-professional: Working together for health and welfare.* London: Routledge.

Letourneux, V. (1998). *Precarious employment and working conditions in the European Union.* Dublin: European Foundation for the Improvement of Living and Working Conditions.

Limborg, H. J. (2001, spring). The professional working environment consultant: a new actor in the health and safety arena. *Human Factors and Ergonomics in Manufacturing & Service Industries, 11*(2), 159-172.

Lorimer, W. & Manion, J. (1996, spring). Team-based organizations: leading the essential transformation. *Patient Focused Care Association Review, 15*(9).

Mearns, K., & Yule, S. (2009, July). The role of national culture in determine safety performance: challenges for the global oil and gas industry. *Safety Science, 47*(6), 777-785.

National Institute for Occupational Safety and Health. (2011). *R2P: research to practice at NIOSH.* Retrieved from http://www.cdc.gov/niosh/r2p/

National Research Council. (2000). *Safe work in the 21st Century: Education and training needs for the next decade's occupational safety and health personnel*. Washington, DC: The National Academies Press.

Norman, D. A. (2010, July/August). The research-practice gap: the need for translational developers. *Interactions, 17*(4), 9-12.

Okafor, E. E. (2007). Globalization, changes and strategies for managing workers: resistance in work organizations in Nigeria. *Journal of Human Ecology, 22*(2), 159-169.

Oxford dictionaries (2014). Oxford: Oxford University Press. Retrieved from http://www.oxforddictionaries.com/us/

Papadopoulos, G., Georgiadou, P., Papazoglou, C., & Michaliou, K. (2010, October). Occupational and public health and safety in a changing work environment: an integrated approach for risk assessment and prevention. *Safety Science, 48*(8), 943-949.

Rosen, M. A., Caravanos, J., Milek, D., & Udasin, I. (2011, July). An innovative approach to interdisciplinary occupational safety and health education. *American Journal of Industrial Medicine, 54*(7), 515-520.

Rosenfield, P. L. (1992, December). The potential of transdisciplinary research for sustaining and extending linkages between the health and social sciences. *Social Science & Medicine, 35*(11), 1343-1357.

Rosenheck, R. A. (2001, December). Organizational process: a missing link between research and practice. *Psychiatric Services, 52*(12), 1607-1612.

Salazar, M. K. (2002, November). Applying research to practice: practical guidelines for occupational health nurses. *American Association of Occupational Health Nurses Journal, 50*(11), 520-525.

Sassen, S. (1998). *Globalization and its discontents*. New York: The New York Press.

Sauter, S., Rosenstock, L. (2000). An American perspective. *Magazine of the European Agency for Safety and Health at Work, 2*, 19-21.

Sorensen, G., & Barbeau, E. M. (2004, October). *Steps to a healthier us workforce: Integrating occupational health and safety and worksite health promotion: state of the science*. Washington, DC. Retrieved from http://www.saif.com/news/CSR_Report/_media/CNSteps.pdf

Sorensen, G., & Barbeau, E. M. (2006, March/April). Integrating occupational health, safety and worksite health promotion: opportunities for research and practice. *La Medicina del Lavoro, 97*(2), 240-257.

Stellman, J. M. (Ed.). (1998). *Encyclopaedia of Occupational Health and Safety*. Geneva: International Labour Office.

Stokes, D. E. (1997). *Pasteur's quadrant: Basic science and technological innovation*. Washington, DC. The Brookings Institution.

Storrie, D. (2002). *Temporary agency work in the European Union*. Luxembourg: Office for Official Publications of the European Communities. Retrieved from http://www.eurofound.europa.eu/pubdocs/2002/02/en/1/ef0202en.pdf

Teusner, A. (2010). "Being" versus "going" native: an account from the OHS field. *Journal of Health & Safety, Research & Practice, 2*(2), 23-33.

Waterman, R. H, Waterman, J. A. & Collard, B. A. (1994). *Toward a career-resilient workforce*. Retrieved from http://hbr.org/1994/07/toward-a-career-resilient-workforce/ib

Wilpert, B. (2009, July). Impact of globalization on human work. *Safety Science, 47*(6), 727-732.

Woods, D. D. (2006, December). Resilience engineering: redefining the culture of safety and risk management. *Human Factors and Ergonomics Society, 49*(12), 1-3.

INDEX

A

Automation　3, 6, 9, 20, 35, 128
Autonomy　3, 4, 8–10, 13, 14, 18, 22, 28, 92, 107

B

Backdrop　32, 33
Behavioral skills　117
Biochips　73, 74
Biotechnologies　35, 37, 47, 49, 60, 73, 100
Bloodstream　53
Bonuses　15, 24
Bridging knowledge　118
Bullying　3, 18, 19

C

Capital contradictions　32
Capitalism　26, 32, 37, 71, 72
Capitalists　65, 71, 72
Capitalist societies　36, 59
Capital power constrains workers organization　72
Capital reproduction　59, 60, 72, 75, 80
Capital reproduction processes　75
Carbon nanotubes　73, 74
Changing world of work　87-90, 94, 95, 110, 129
Civil society　62, 69
Collective bargaining agreement　64, 65
Collins dictionary　112
Commitment　7, 9, 87, 117, 126
　joint　132
　personal　116, 117

Marcela G. Ribeiro (Ed)
All rights reserved-© 2014 Bentham Science Publishers

knowledge/non-knowledge   71
  science of   51, 60, 72
Production chains   15, 23, 24
Production processes   6-8, 18, 56, 63, 70
  automation of   3, 9
Production rates   14
Production rationales, new   9
Production systems   7, 9, 10, 24
Productive rationality, new   20, 22
Products, nanostructured   65
Properties, health-giving   130
Public resources, allocation of   58, 60

Q

Qualification, formal   11
Quality OHS research outputs, high   120

R

Raw materials, natural   75
Recognition skills   113
Redistribution   58
Regarding nanotechnology development   69
Regarding nanotechnology regulation   63
Regarding nanotechnology research policies   70
Regarding new health   37
Regarding new technologies   65, 79
Research
  biotechnological   77
  practical application of   121
Research outcomes, scientific   123
Resources
  human   68, 94, 95, 121
  natural   75